Biology Workbook
Middle School Science
Volume 1

AF881068

Emma McErlane

© Emma The Teachie Limited 2025-Present. All rights reserved.

© Emma The Teachie Limited 2025. All rights reserved.

All works created by Emma The Teachie Limited. All rights reserved. No part of this publication may be reproduced, distributed, or transmitted in any form or by any means. This includes photocopying, recording, or other electronic or mechanical methods without prior permission of the publisher, except in the case of brief quotations embodied in critical reviews and other noncommercial uses permitted by copyright law.

No part of this product may be used or reproduced for commercial use.

Contact the author:
Emmatheteachie.com

 © Emma The Teachie Limited 2025-Present. All rights reserved.

CONTENTS

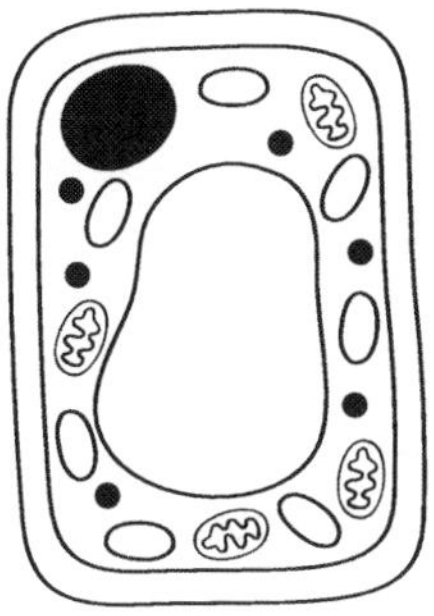

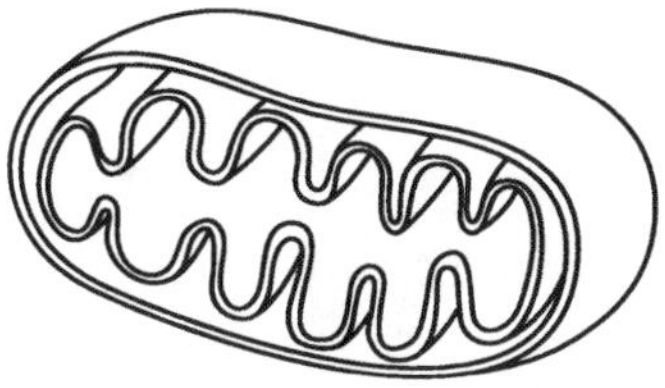

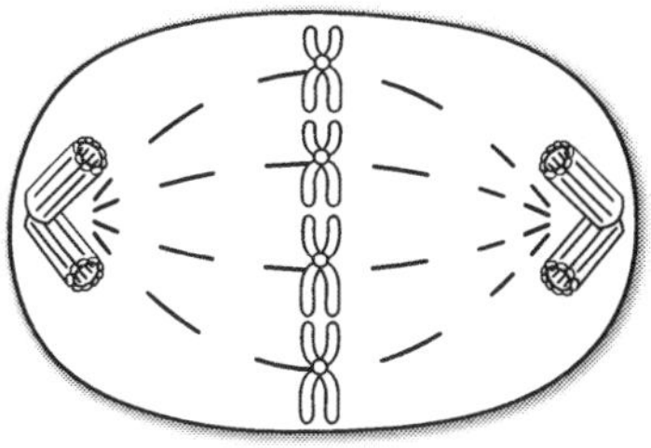

© Emma The Teachie Limited 2025-Present. All rights reserved.

How to use this workbook

A note for parents:

This workbook is designed to be rigorous yet FUN for your child!

These are standalone lessons that they can complete on their own. You do not need to purchase a supplemental textbook. All of the lessons come with text passages that explain the key concepts and vocabulary.

These are custom illustrated so the content is interesting and age appropriate.

Every activity and quiz has an answer key. These are found in the back of the workbook.

You can decide whether to remove these (cut them out), or leave them in for your child to self-check their work. Self-checking can be a wonderful way for students to reflect on their understanding and see where they can improve. Alternatively, you can check their work together.

The materials your child will need for this workbook are:

- Pens (ideally two colors — one color for completing the activities, one color for adding corrections)
- Color pencils and a normal pencil
- Highlighter (they can also use a color pencil)
- Ruler

Biology is a very visual subject, and students are encouraged to add their own doodles, color in, and complete diagrams, storyboards, etc.

I hope you and your child both enjoy using this workbook. *And if you do, please leave a review!*

A note for students:

By the end of this workbook, you will feel like a genius!

You'll have learned about what is happening inside your cells and other cells in our world.

Biology is a fascinating subject, and there is always more to discover and learn. Hopefully these lessons ignite your curiosity!

The next two pages explain how to use this workbook. Each lesson follows the same format: Process → Activate → Extend → Review → Quiz.

Make this workbook your own: color in the pictures (when you need a brain break!), add your own doodles and speech bubbles, write notes in the margins, and reread the text passages.

The more you put into your learning, the more you will get out of it!

© Emma The Teachie Limited 2025-Present. All rights reserved.

PROCESS

Each lesson begins with two pages of text to read (except cell organelles, it has 4 pages — there are a lot of organelles!).

It's really important to process what we read.

To help do this, there are tasks to complete after reading:

- highlighting new keywords;
- underlining definitions;
- squiggly underlining old keywords; ← *to link to your prior knowledge!*
- adding relevant doodles. ← *to make it memorable!*

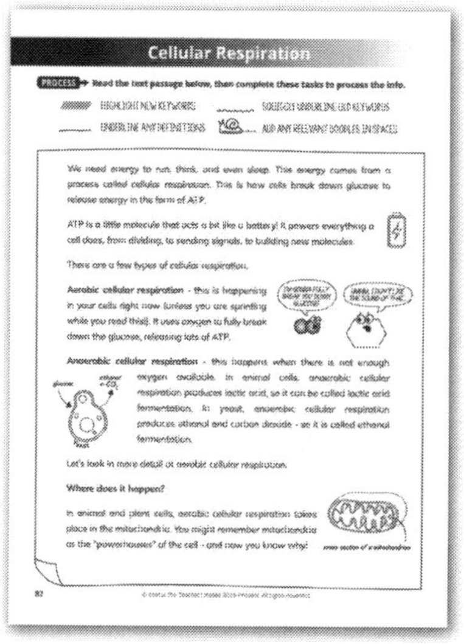

ACTIVATE

You'll apply what you've learned in a series of activities.

There are all kinds of activities, including match-ups, labeling, storyboards, and Venn diagrams.

Occasionally, there are "TAKE IT FURTHER" tasks to stretch your understanding!

You can go back to the text passage at any time.

Unless the instructions tell you to use it, you don't need the internet to complete the activities.

↑ *Everything you need to know is already in the text passage, the activity, or your brain!*

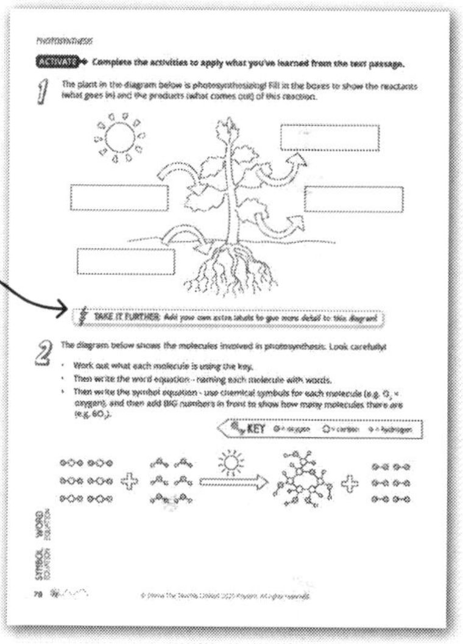

EXTEND

The extend task goes beyond the content in the text passage. It may give you new information or require you to think creatively. Don't worry if you feel like you don't know the answer!

Science is always changing as new evidence emerges. So let's get comfortable with being challenged by new info and being asked deeper questions!

REVIEW

Each lesson has a review to check what you've understood.

To start, try and complete it without looking back at previous activities or the text passage. After you've done as much as you can, then you can look back if you need to.

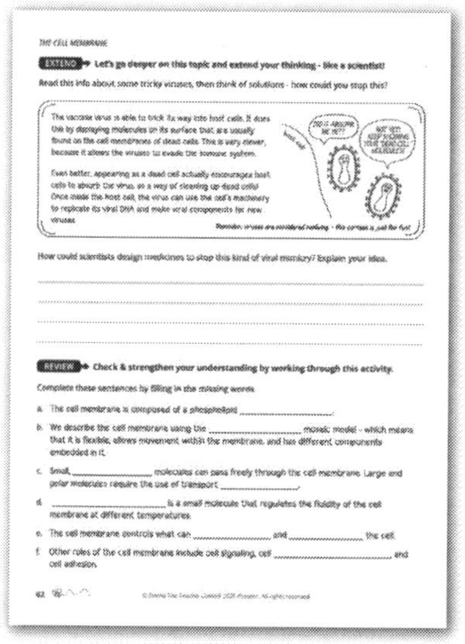

© Emma The Teachie Limited 2025-Present. All rights reserved.

Before you take the quiz...

Use the lesson answer key at the back of the workbook to check your answers to the activate, extend, and review activities. Make corrections to any you got wrong and add in extra details if you missed parts.

Use a different color pen to make corrections stand out!

This will strengthen your understanding before you take the quiz.

QUIZ

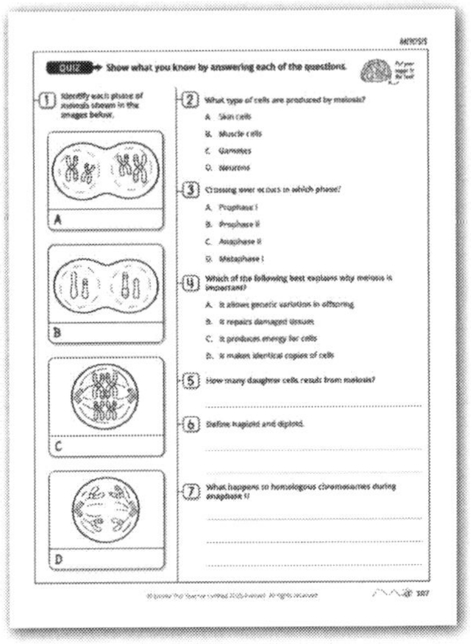

At the end of each lesson, there is a quiz. This is your chance to show what you know!

Take your time and read through the questions carefully. For multiple choice questions, read all of the options before choosing your answer.

There are some short answer questions as well — try to use keywords in your answer.

Once you're finished, use the answer key to mark your quiz. Make corrections for any questions you got wrong.

UNIT REVIEW CROSSWORDS

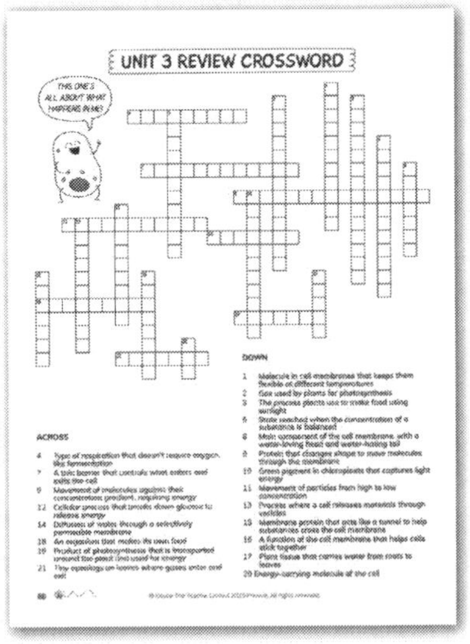

There are four unit review crosswords. These are great for seeing how many keywords and concepts you remember from across the whole unit!

You can do these at the end of each unit, or you can wait until you have completed the entire workbook and use them as a way to prepare for the workbook assessment.

How to complete the crosswords:

"Across" clues are for words written horizontally.

"Down" clues are for words written vertically.

Sometimes the answer is two words. Do **not** include a space between the words when writing your answer into the crossword.

WORKBOOK ASSESSMENT

Instructions for the assessment are given at the start of it. Do the assessment when you have finished the entire workbook!

© Emma The Teachie Limited 2025-Present. All rights reserved.

UNIT 1 THE STUDY OF LIFE

© Emma The Teachie Limited 2025-Present. All rights reserved.

The Scientific Method

PROCESS → **Read the text passage below, then complete these tasks to process the info.**

HIGHLIGHT NEW KEYWORDS

SQUIGGLY UNDERLINE OLD KEYWORDS

UNDERLINE ANY DEFINITIONS

ADD ANY RELEVANT DOODLES IN SPACES

The scientific method is used to discover the answers to questions about the world and how it works. It is a process that is used by scientists in many different fields!

Let's take a look at the steps of the scientific method. Note that these are non-linear, meaning that it doesn't always happen in this exact order.

Observation: We observe the world and how it works with our senses. You may notice something that catches your attention. This is an observation.

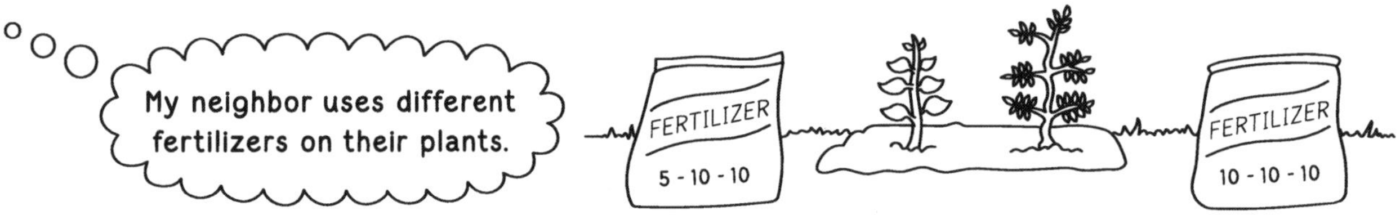

Question: Based on your observation, you can ask a question. A scientific question must be testable and not based on opinion. Example:

Q **Does the percentage of nitrogen in fertilizer affect plant growth?**

Research: Once you have a question, you need to carry out research. Find out what is currently known on the topic and if the question has already been answered. If it has already been answered, you can ask another question.

Fertilizers contain different percentages of nitrogen and other micronutrients. Nitrogen is used for the production of chlorophyll and amino acids, both of which are needed for plant growth.

Hypothesis: Now you create a hypothesis. A hypothesis is a prediction about

© Emma The Teachie Limited 2025-Present. All rights reserved.

the answer to your question. This is based on your research — it is not opinion based. We use "if . . . then . . . because . . ." statements to help write a strong hypothesis.

If a fertilizer has a higher percentage of nitrogen, then plant growth is increased because nitrogen is an important nutrient for plants.

Experiment: You need to design an experiment to test your hypothesis. It needs to have an independent variable, a dependent variable, and control variables. It should be repeatable by you and by other scientists.

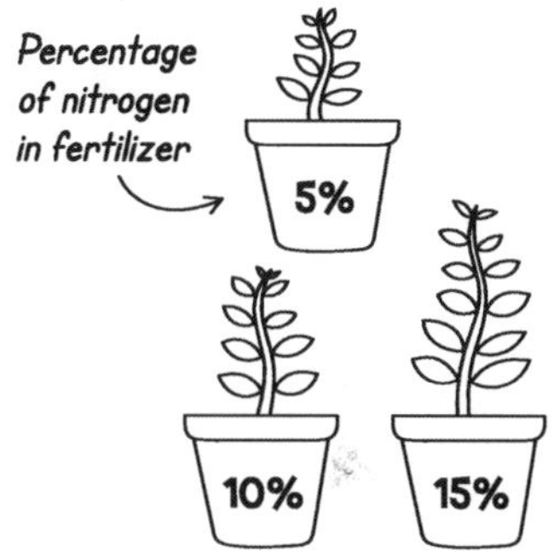

- Independent variable: This is what you **change**.
- Dependent variable: This is what you **measure**.
- Control variables: These are what you **keep constant**.

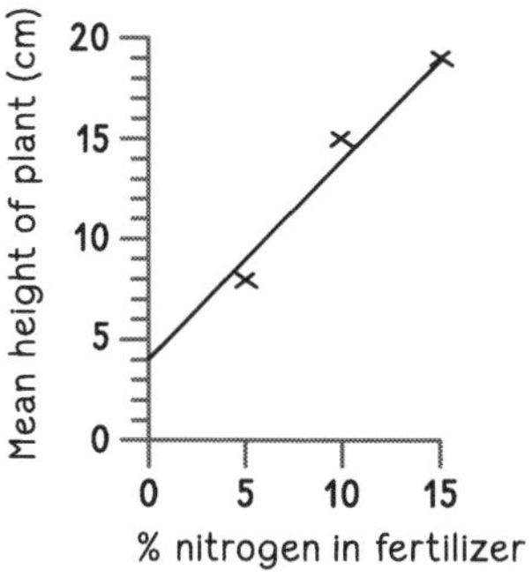

Analysis: Once you have conducted your experiment and collected data from it, you need to analyze the data. This can include completing calculations and drawing graphs and charts. This can help you to identify any trends in the data and spot any errors.

Conclusion: This explains what your data shows and whether it supports your hypothesis. A clear way to do this is with Claim-Evidence-Reasoning: The claim is your answer, the evidence is your data, and the reasoning explains why the data supports it. If unsupported, form and test a new hypothesis.

Complete the conclusion

The higher the percentage of nitrogen in fertilizer... ____________
This supports / rejects my hypothesis. The plant with 15% nitrogen grew the most, and the plant with 5% nitrogen grew the least. Plants use nitrogen for cell growth.

Report: Share your findings with others! Scientists share their results in presentations and publish them in scientific journals. These are peer-reviewed before publication, which means other scientists check the results are repeatable and of a good scientific standard.

© Emma The Teachie Limited 2025-Present. All rights reserved.

ACTIVATE → Complete the activities to apply what you've learned from the text passage.

1 Draw lines to match each step of the scientific method to its correct description. Then, number each step to show the order they are most commonly carried out.

(Remember, they can loop back & repeat!)

Step		Description
☐ EXPERIMENT	A	A prediction written as an "if... then... because..." statement, based on research.
☐ HYPOTHESIS	B	Collecting information that is already known to learn more about the topic.
☐ CONCLUSION	C	Looking at the world with your senses and noticing something interesting.
☐ RESEARCH	D	Explaining whether the data supports your hypothesis or not.
☐ REPORT	E	Testing the hypothesis by changing one variable, measuring another, and keeping the rest the same.
☐ ANALYSIS	F	Looking at the data, making graphs, and identifying patterns or errors.
☐ QUESTION	G	Asking a testable, scientific question that is not based on opinion.
☐ OBSERVATION	H	Sharing results with other scientists, usually through journals or presentations.

2 It can be tricky to remember which variable is which — practice makes this a lot easier. For each experiment, name the variables: independent variable (IV), dependent variable (DV), and control variable(s) (CV). The first one is an example.

Gloria releases a toy car from ramps that are different heights. She measures the distance the car travels from the bottom of the ramp.

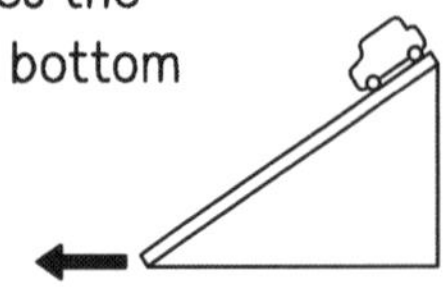

IV: Height of the ramp

DV: Distance traveled from the ramp's end

CV: Ramp length, surface material of ramp, type of toy car, where the car is released from.

Josh plants seeds in sand, clay, and garden soil. He waters each with 200 mL of water and records how much drains through the soil after 10 minutes.

IV: ______________________

DV: ______________________

CV: ______________________

____

© Emma The Teachie Limited 2025-Present. All rights reserved.

Alexis measures her heart rate after walking, jogging, and running quickly for 2 minutes each. She records her beats per minute right after each activity, and rests for 10 minutes between activities.

IV:

DV:

CV:

Luca counts the number times a cricket chirps in 15 seconds in 15°C, 20°C, and 25°C rooms. He gives the cricket 30 minutes to adjust to each new temperature before counting.

IV:

DV:

CV:

TAKE IT FURTHER: Think of a hypothesis for each experiment, using "if...then...because..." statements.

3

Here is the data collected for two experiments. Write Claim-Evidence-Reason (CER) statements based on the results. You may need to do some research for the reasoning part.

ALEXIS

Type of exercise	Heart rate (beats per minute)
Walking	85
Jogging	126
Running	184

LUCA

Temperature (°C)	Number of chirps (per minute)
15	20
20	28
25	39

Look up "Dolbear's law" – it's fascinating!

EXAMPLE FOR GLORIA: The taller the ramp, the further the car travels from the end of the ramp. A 5cm ramp made the car travel 7cm, while a 10cm ramp made it travel 11cm. A higher ramp gives the car more gravitational potential energy, which gives it more kinetic energy, so it travels farther.

Alexis:

Luca:

© Emma The Teachie Limited 2025-Present. All rights reserved.

EXTEND ➜ Let's go deeper on this topic and extend your thinking — like a scientist!

Think of a problem in your daily life, and come up with a question about it. For example: "Which snack keeps me full the longest?" Write out how you could use the scientific method to test it.

Remind yourself: What kind of structure do I need to use for the hypothesis?

YOUR QUESTION!

HYPOTHESIS

EXPERIMENT
{ *Include the variables here* }

REVIEW ➜ Check & strengthen your understanding by working through this activity.

Complete these sentences by filling in the missing words.

a. An observation is taking in information using our ____________________.

b. A scientific question must be ____________________ and not based on personal opinion.

c. Research can reveal if a question has already been ____________________ and help in planning an experiment.

d. Before starting an experiment, scientists make a ____________________ - this is a testable answer to the question. We can use "if... ____________________... because..." statements for these.

e. The experiment tests the hypothesis. It has an ____________________ variable (what we change), a ____________________ variable (what we measure), and a number of ____________________ variables (what we keep the same each time).

f. After the experiment, it is important to ____________________ the data that was collected. This can identify ____________________ in the data.

g. Then, scientists write a ____________________ to explain what their experimental data showed and why. This can be done using ____________________ statements.

h. Finally, scientists share their results with their peers (other scientists) by giving presentations and writing ____________________ that get published in scientific ____________________.

© Emma The Teachie Limited 2025-Present. All rights reserved.

QUIZ → **Show what you know by answering each of the questions.**

1 Identify each type of variable from this experiment:

Lila stores slices of bread in these places: a sealed bag, on the counter, and in the refrigerator.

After one week, she measures the percentage cover of mold on the top of each slice.

She makes sure she wears a face mask and gloves while handling the moldy bread.

% cover of mold

A

Type of bread

B

Location of the bread

C

2 Which of these is the correct definition of a hypothesis?

A. An opinion about how the world works

B. A prediction based on research, testable by experiments

C. A fact proven by scientists

D. A list of control variables

3 What is the purpose of analyzing data?

A. To make the hypothesis

B. To share findings with others

C. To find patterns and trends

D. To write research notes

4 Why are scientific journals peer-reviewed?

A. To make them easier to read

B. To check the work meets scientific standards

C. To keep results secret

D. To write more hypotheses

5 Define “observation” in your own words.

____

____

6 What structure should we use when writing a clear conclusion?

____

7 Write a hypothesis in “if... then... because...” format about how the amount of light might affect plant growth.

____

____

____

© Emma The Teachie Limited 2025-Present. All rights reserved.

Characteristics of Life

PROCESS → **Read the text passage below, then complete these tasks to process the info.**

HIGHLIGHT NEW KEYWORDS

SQUIGGLY UNDERLINE OLD KEYWORDS

UNDERLINE ANY DEFINITIONS

ADD ANY RELEVANT DOODLES IN SPACES

Biology is the study of living things — also known as organisms. So, it is important that we can define what a living thing is!

THINK: What do all living things have in common? Including plants, bacteria, butterflies, and you!

We call these the characteristics of life. Let's see how many you thought of.

1. Cells - Organisms are made of cells. Sometimes it is a single cell - we call these organisms unicellular. And sometimes it is many cells - we call these organisms multicellular. In multicellular organisms, the cells are highly organized (more on this in the next lesson).

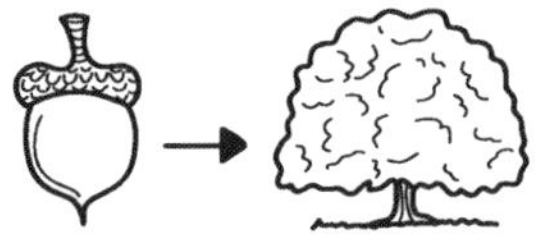

2. Growth - Organisms are capable of growing larger. You grew from a single fertilized egg cell into the person you are today! Even single-celled organisms like bacteria are capable of growth.

3. Response to stimuli - Organisms can respond to things in their internal and external environments. We call these stimuli. For example, plants grow towards sunlight. In this situation, the sunlight is the stimulus.

4. Reproduction - Organisms can reproduce to create more organisms. Reproduction can be sexual or asexual.

© Emma The Teachie Limited 2025-Present. All rights reserved.

5. Homeostasis - Organisms control their internal environment, keeping it in a stable range. We call this homeostasis. For example, we control our body temperature by sweating, shivering, and other responses.

6. Metabolism - Organisms take in and use energy for chemical reactions to keep it alive. This is known as metabolism. The energy can be used for growth, movement, and reproduction. Organisms get the energy in different ways: plants get energy from light and animals get energy from eating other organisms.

7. Evolution - Organisms can adapt to their external environment and evolve because of it. This is possible through the passing on of genetic material, called DNA, from one generation to the next.

If something is missing one or more of these characteristics, then it is not alive.*

For example, a strawberry on your plate is not alive. It is not able to reproduce to create a new strawberry plant, even though it has DNA inside its cells.

**There are a few exceptions, which you'll explore in the activities!*

THINK: Can you think of any exceptions yourself?

© Emma The Teachie Limited 2025-Present. All rights reserved.

ACTIVATE → **Complete the activities to apply what you've learned from the text passage.**

1 Put each of these organisms into the correct column in the table to show if it is unicellular or multicellular.

- Human
- Bacterium
- Cat
- Amoeba
- Tree
- Paramecium

Unicellular	Multicellular

TAKE IT FURTHER: Come up with one more to add to each column!

2 Name each characteristic of life using the descriptions... but wait! One description is missing! Write your own description for the last characteristic of life.

NAME	DESCRIPTION
	Living things become larger in size over time, for example, the growth of a bacterial cell or a puppy.
	Maintaining a stable internal environment, for example, body temperature, to allow an organism's cells to function.
	Living things need energy to carry out chemical reactions that keep it alive.
	Living things respond to internal and external cues, for example, quickly removing your hand from a hot surface.
	Living things are made of cells. They can be unicellular or multicellular.
	Living things have characteristics that help them survive & reproduce in their environment. These can change over time.
	______________________________

© Emma The Teachie Limited 2025-Present. All rights reserved.

3 Imagine you are explaining to a younger student why a robot is not alive. Write 3–5 sentences, using at least three characteristics of life to support your explanation.

4 For each item in the list, decide if it is living or nonliving and explain why you think that.

TOP TIP: If you think it is nonliving, which characteristic(s) of life is it missing?

	LIVING	NONLIVING	EXPLANATION
Rock	☐	☐	
Seaweed	☐	☐	
Snowflake	☐	☐	
Wooden table	☐	☐	
Grasshopper	☐	☐	

5 Have you heard of a zonkey? Read this information and then decide... is a zonkey alive?

Zonkeys are the offspring of a male donkey and a female zebra. They have characteristics of both parents, as you can see in the picture! Zonkeys can sometimes suffer from health problems, and they are sterile (infertile) due to the odd number of chromosomes they inherited from their parents. They can live for around 15–25 years, which is similar to a wild zebra. They are herbivores, like their parents.

I think zonkeys are...

Give reasons to support what you think!

© Emma The Teachie Limited 2025-Present. All rights reserved.

EXTEND → **Let's go deeper on this topic and extend your thinking — like a scientist!**

Viruses are tiny particles that can infect body cells. They have their own genetic material (DNA or RNA), but they cannot reproduce on their own or carry out metabolism; that's why they need host cells. They are capable of evolving, e.g., to evade the immune system.

"Viruses are alive." Give arguments for and against this statement. Then, explain what YOU think!

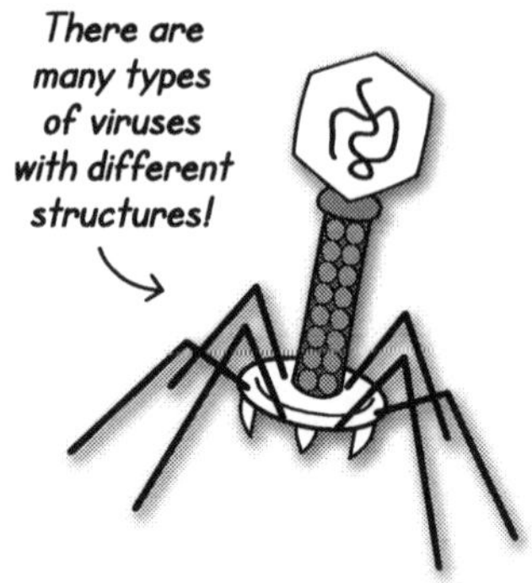

REVIEW → **Check & strengthen your understanding by working through this activity.**

Write all of the different characteristics of life around the bubble. Then, add a doodle for each one to help you remember what it is!

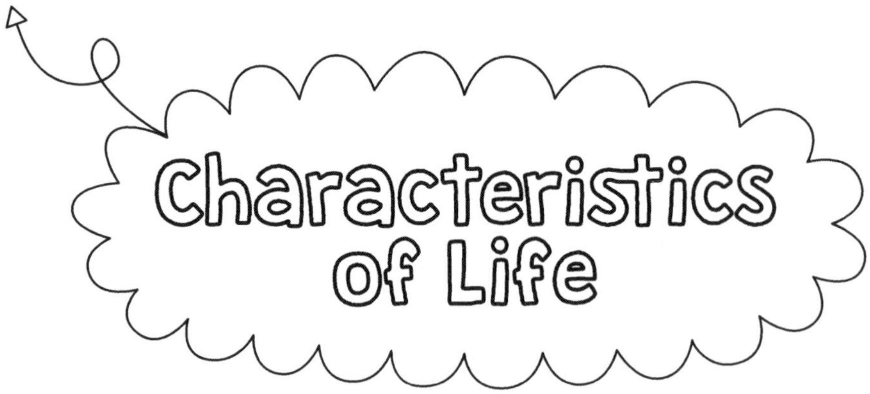

© Emma The Teachie Limited 2025-Present. All rights reserved.

QUIZ ➔ **Show what you know by answering each of the questions.**

1 Identify the characteristic of life represented by each picture.

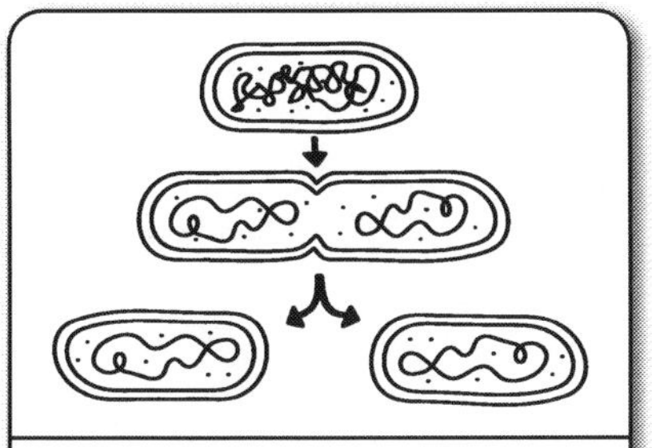

A

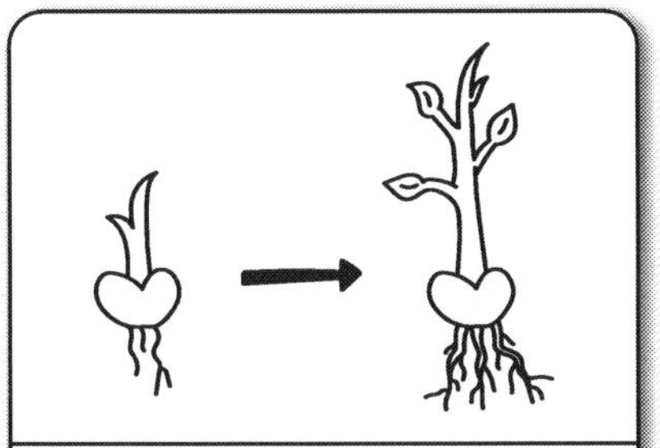

B

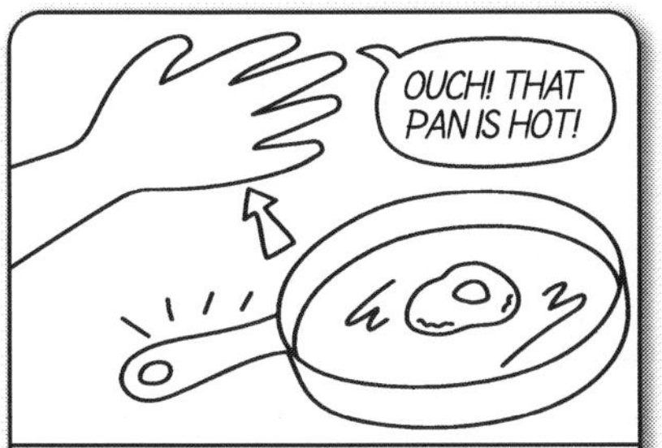

C

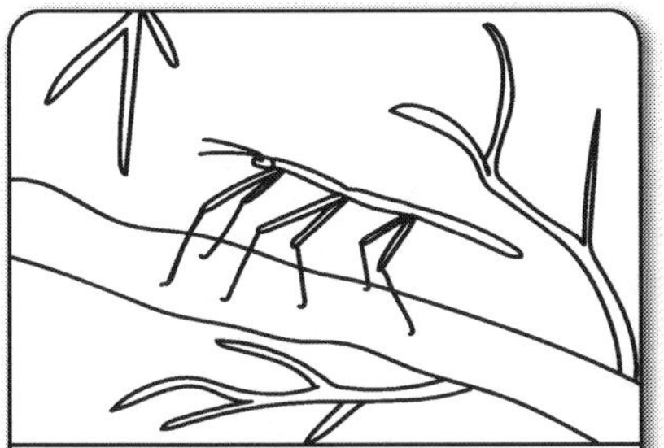

D

2 Which of these is NOT a characteristic of life?

A. Breathing

B. Having cell(s)

C. Homeostasis

D. Reproduction

3 Which of the following is an example of a response to stimuli?

A. A cactus storing water in its stem

B. A dog panting when it is hot

C. A mushroom using energy from decaying matter

D. A sunflower having DNA

4 Why is a cherry on a plate not alive?

A. It no longer has DNA

B. It cannot reproduce

C. It does not have cells

D. It cannot ripen

5 Which of the following best explains what metabolism is?

A. Producing offspring

B. Maintaining a stable body temperature

C. Using energy for essential chemical reactions

D. Changing characteristics over generations

6 Organisms are always made of one or more cells. Single-celled organisms are called ____________________, whereas organisms made from many cells are called ______________.

7 Give one example of homeostasis in humans.

____

© Emma The Teachie Limited 2025-Present. All rights reserved.

Levels of Organization

PROCESS ➜ **Read the text passage below, then complete these tasks to process the info.**

HIGHLIGHT NEW KEYWORDS — SQUIGGLY UNDERLINE OLD KEYWORDS

UNDERLINE ANY DEFINITIONS — ADD RELEVANT DOODLES ON THE SIDES

Living things are highly organized, and it is no surprise that there is an order to it. Let's take a look at the levels of organization.

The smallest living level of organization is the cell.

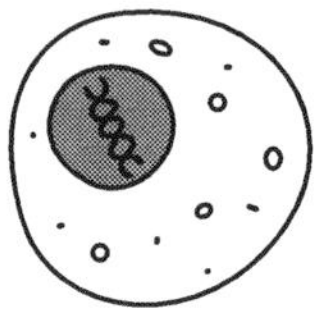

(Notice how I said the smallest ***living*** *level? That's because even cells are made up of smaller units, like organelles, and smaller than that, atoms — however organelles and atoms are not actually alive).*

Every living thing starts with a cell, from bacteria and protists to animals and plants.

When lots of cells work together to perform a specific function, we call this a tissue. Connective tissue and muscle tissue are two examples.

Then we've got organs — these are made up of different tissues working together. The lungs, stomach, and brain are all organs.

When multiple organs work together to carry out a function, such as digestion, we call it an organ system. You can probably think of a few different organ systems in your body!

THINK: How many organ systems can you name?

Last up on an individual level is the actual organism. This is a living thing — like you, a pumpkin, or a cat.

© Emma The Teachie Limited 2025-Present. All rights reserved.

Note: Not all organisms contain all of these levels. Many organisms are single-celled, like bacteria and amoeba.

Beyond Organisms

We can go beyond the levels of organization in individuals, and look at ecological levels of organization.

When there are multiple organisms of the same species living in the same area at the same time, this is called a population. Example: A herd of gazelle is a population.

A community is all of the populations of different species in a given area at the same time.

Then we have ecosystems. An ecosystem is all of the biotic (living) and abiotic (nonliving) factors in an area. So basically, an ecosystem is the community and its physical environment.

Lastly, we have the biosphere. This is all of the parts of the Earth which contain living organisms. It encompasses all of the ecosystems on the planet!

© Emma The Teachie Limited 2025-Present. All rights reserved.

ACTIVATE → Complete the activities to apply what you've learned from the text passage.

1 Match each level of organization to its description. Then, number them to show the order of the levels of organization, with 1 being the smallest level of organization.

Add the numbers here

- ☐ Organ
- ☐ Tissue
- ☐ Cell
- ☐ Organ system
- ☐ Organism

- **A** The basic unit of structure and function in living organisms.
- **B** A group of organs with related functions, working together to perform body functions.
- **C** A group of tissues that work together to perform a specific function.
- **D** A living thing that carries out all basic functions for life.
- **E** A group of similar cells that work together to perform a specific function.

2 Decide which level of organization is represented by each picture. Some levels may be shown more than once!

A.

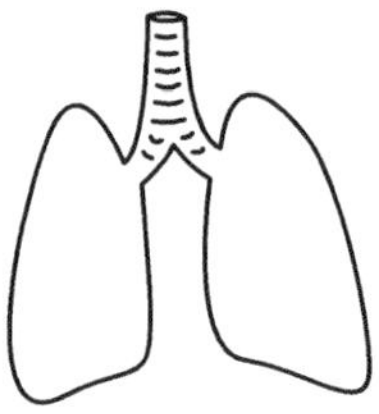

____

B.

____

C.

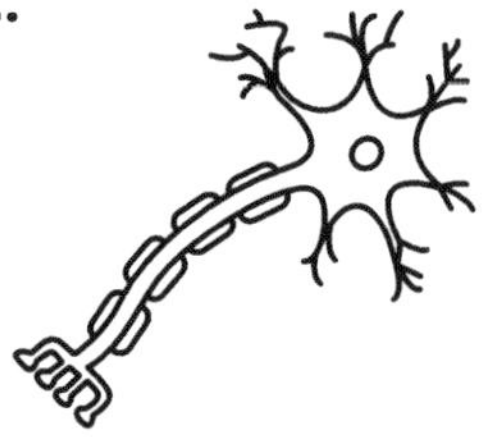

____

D.

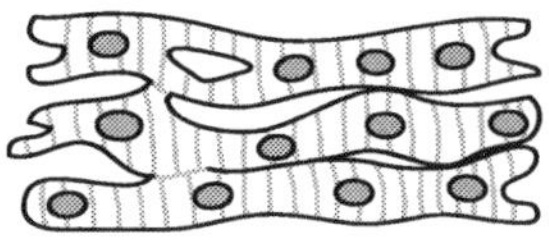

____

E.

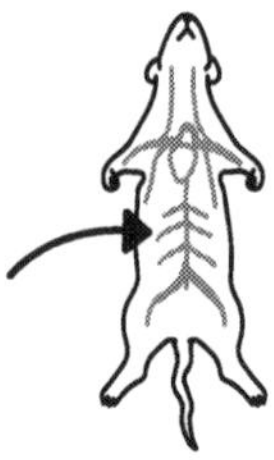

____

F.

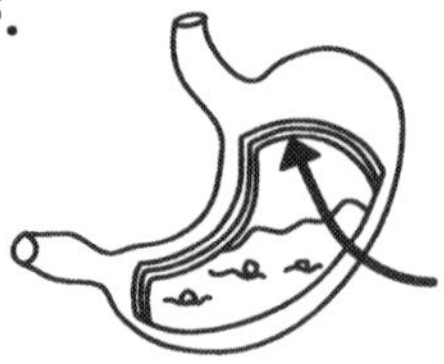

____

G.

____

H.

____

⚡ TAKE IT FURTHER: How many do you know the name of?

Write these below each level!

© Emma The Teachie Limited 2025-Present. All rights reserved.

3 In the space below, doodle some of the organs found in your body! Then, name each one.

TAKE IT FURTHER: Which organ system does each organ below to?

4 Below is a description of an organ system found in plants. It describes each level of organization within that organ system. Circle the examples of each level of organization.

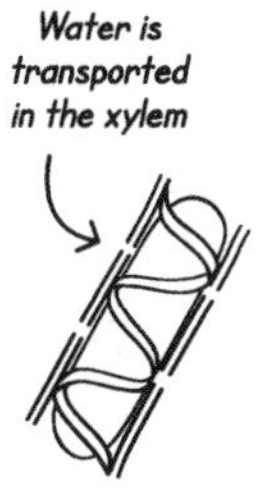

Many plants, like ferns, trees, and flowering plants, have a vascular system. Its function is to move water, nutrients, and minerals around the plant. The organs of this organ system are the stem, roots, and leaves. The tissues are the xylem and phloem. The xylem is for transporting water and minerals, while the phloem is for transporting sugar and other organic molecules. And finally, the cells of the vascular system are those found within the xylem and phloem – these include tracheids, vessel elements, sieve tube elements, phloem parenchyma, and companion cells.

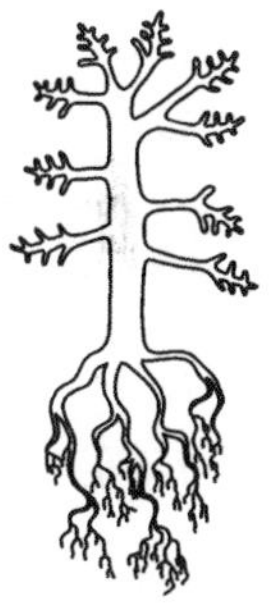

Now, it's your turn. Choose an organ system in **the human body**. In full sentences, describe each level of organization found within it. You can include multiple examples. You may need to do some research for this. (Set a timer — maximum 3 minutes!)

© Emma The Teachie Limited 2025-Present. All rights reserved.

EXTEND ➜ **Let's go deeper on this topic and extend your thinking - like a scientist!**

Pick a body system — now imagine that one organ or one tissue in that body system is damaged or failing. E.g., lungs damaged by smoke or eyesight worsening as a result of severe diabetes.

- How would this affect the functioning of the organ system? How would it affect the organism?
- How would the body adapt to this? Are there any medical interventions that could help?

DAMAGED ORGAN/TISSUE:

IMPACT OF DAMAGE ON ORGAN SYSTEM & ORGANISM:

BODY'S RESPONSE + POSSIBLE MEDICAL INTERVENTION:

REVIEW ➜ **Check & strengthen your understanding by working through this activity.**

Answer the clues to complete the crossword puzzle. Reminder: if the answer is two words, do not leave a space between the words.

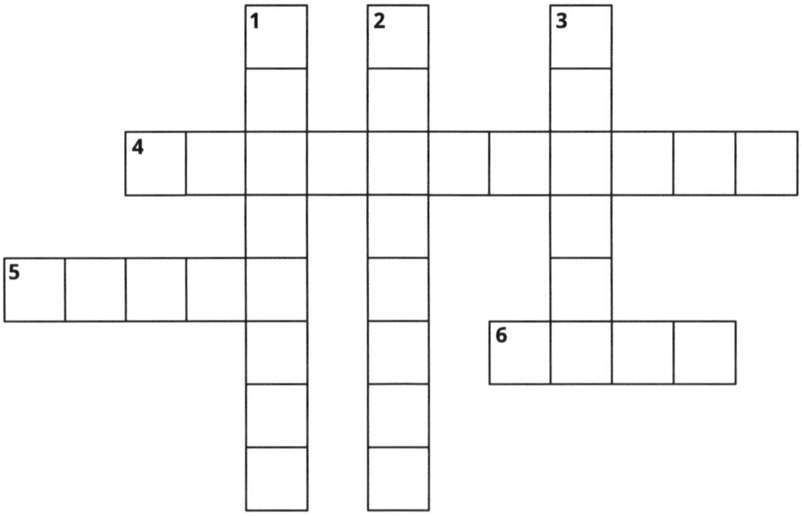

ACROSS
4 A group of organs with related functions, working together to perform body functions.
5 A group of tissues that work together to perform a specific function.
6 The basic unit of structure and function in living organisms.

DOWN
1 A living thing that carries out all basic functions for life.
2 This word is another way of saying "job" — what the object does.
3 A group of similar cells that work together to perform a specific function.

© Emma The Teachie Limited 2025-Present. All rights reserved.

TEST ➔ **Show what you know by answering each of the questions.**

1 Identify each level of organization shown in the images.

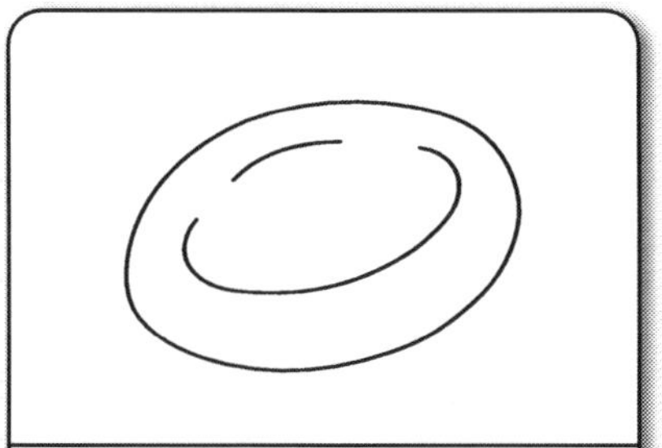

A

B

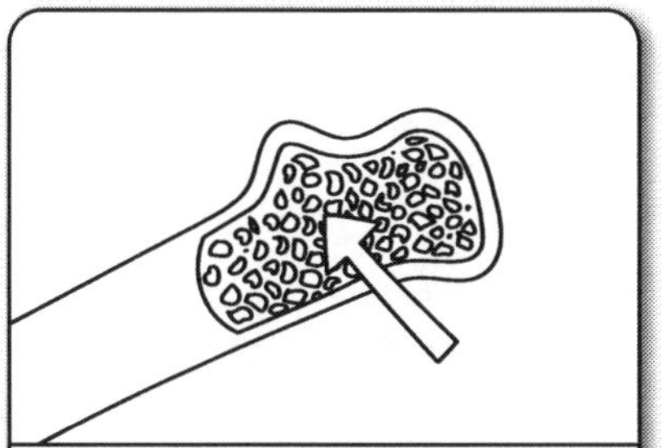

C

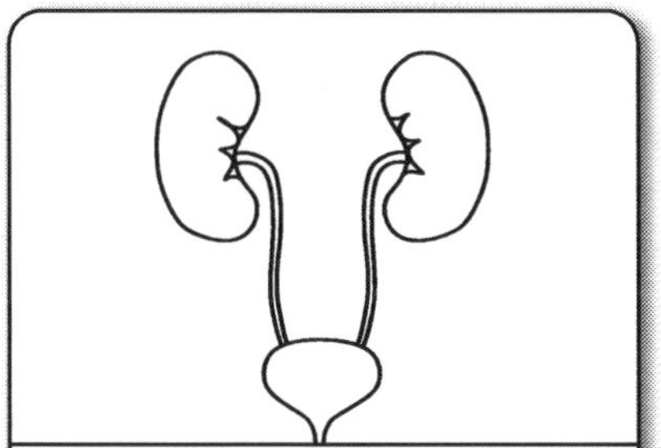

D

2 As we move from the smallest level of organization to the largest, what changes?

A. Numbers

B. Intelligence

C. Adaptation

D. Complexity

3 Which of these is the definition for an organ?

A. The basic unit of structure and function

B. A collection of organ systems that work together to perform the same function

C. A collection of tissues that work together to perform a specific function

D. A collection of cells that work together to perform a similar function

4 Which is the correct order for the levels of organization in living things?

A. Organism → Organ → Organ system → Tissue → Cell

B. Organism → Organ → Organ system → Cell → Tissue

C. Organism → Organ system → Organ → Tissue → Cell

D. Organ → Organism → Organ system → Tissue → Cell

5 Which level of organization does the heart fall into?

____

6 Organelles are not usually included in the hierarchy of organization in living things because they are not... (complete this sentence).

____

7 True or false: Amoeba are unicellular and do not have levels of organization beyond the cell. This means they are not organisms.

____

© Emma The Teachie Limited 2025-Present. All rights reserved.

UNIT 1 REVIEW CROSSWORD

ACROSS

5 A group of similar cells working together to do a job
11 The summary of what you learned from an experiment
12 Scientists carry this out to test their hypothesis
13 A possible explanation or predication that can be tested
14 A publication where scientists share their research results
15 A structure made of different tissues that work together
16 The ability of an organism to maintain a steady internal environment
18 All the different populations living together in an area
19 When an organism increases in size or cell number

DOWN

1 All living and nonliving things that interact in an environment
2 Another word for a living thing
3 All the members of one species living in a certain area
4 The process of producing offspring
6 The variable in an experiment that is changed on purpose
7 Information gathered using the senses
8 A nonliving particle that infects and multiplies within living cells
9 All the chemical reactions that happen inside an organism
10 The variable that is measured or observed in an experiment
13 A possible prediction that can be tested
17 A type of variable that remains the same throughout an experiment
18 The basic unit of life

© Emma The Teachie Limited 2025-Present. All rights reserved.

UNIT 2 CELLS AND THEIR STRUCTURES

© Emma The Teachie Limited 2025-Present. All rights reserved.

Cell Theory

PROCESS → **Read the text passage below, then complete these tasks to process the info.**

HIGHLIGHT NEW KEYWORDS — SQUIGGLY UNDERLINE OLD KEYWORDS

UNDERLINE ANY DEFINITIONS — ADD ANY RELEVANT DOODLES IN SPACES

As you just learned, having cells is a characteristic of life — every living thing on this planet has a cell (or many of them). But we didn't always know that cells existed! We are going to take a look at the scientists who discovered cells and the three components of cell theory.

Let's start with the key components of cell theory:

1 - All organisms are composed of one or more cells.

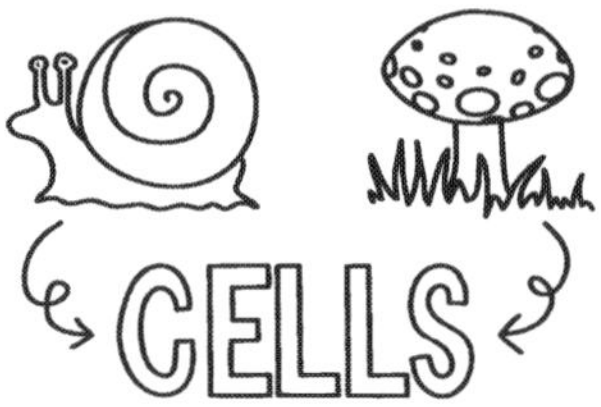

2 - The cell is the basic unit of structure and organization in organisms.

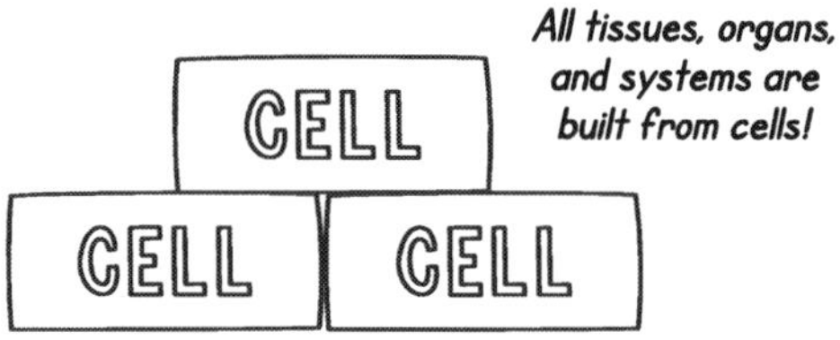

3 - All cells come from preexisting cells.

How did we come to know this? It all started with the development of the microscope.

It is claimed that a spectacle maker called Zacharias Janssen created the first compound microscope in 1590.

Many naturalists and biologists were interested in using these to view the hidden world around them.

In 1665, Robert Hooke used a microscope to view cork. He noticed it was made of

© Emma The Teachie Limited 2025-Present. All rights reserved.

tiny little "blocks". He named these cells, after noting a similarity to monk cells.

THINK: If they weren't already called cells... what would **YOU** call them?

A few years later in the 1670s, Anton Van Leeuwenhoek created and used a more powerful microscope to view what he called "animalcules". He viewed cells including bacteria, red blood cells, sperm cells, and more.

It would be another 150 years before Matthias Schleiden proposed in 1838 that all plants are made up of cells.

Theodor Schwann's announcement soon followed in 1839 that all animals are made of cells. It became clear that all living things are composed of cells.

The final component of cell theory was declared by Rudolf Virchow in 1855. Using the research of others, he confirmed that all cells come from preexisting cells.

Prior to Virchow's theory, many scientists believed in spontaneous generation — the idea that some living things could arise from nonliving matter. While this might sound a little far-fetched to you, science is continually evolving based on the available evidence. In a few hundred years, people in the future may be laughing at some of our beliefs!

© Emma The Teachie Limited 2025-Present. All rights reserved.

ACTIVATE → **Complete the activities to apply what you've learned from the text passage.**

1. State each of the three components of cell theory. Try to do this from memory first, using your own words! Then you can come back and add more details or corrections.

2. Many scientists contributed to our knowledge of cells and cell theory. Draw lines to match each scientist to their contribution.

VIRCHOW	Declared that all animals were made of cells.
JANSSEN	Created the first compound microscope.
LEEUWENHOEK	Invented a more powerful microscope, and used it to observe cells he called "animalcules."
SCHLEIDEN	Viewed cork under a microscope, and named the individual components "cells."
HOOKE	Announced that all plants were made of cells.
SCHWANN	Confirmed that all cells come from preexisting cells.

© Emma The Teachie Limited 2025-Present. All rights reserved.

3 Microscopes were key in the discovery of cells. They have been improved over the centuries. The one commonly used in classrooms is called a compound light microscope. Read the information below, and then use it to label each part of the microscope.

- **Eyepiece** (ocular lens): This is the lens we look through at the top of the microscope. It usually magnifies the image 10 times.
- **Objective lenses**: These are found on a rotating piece called the **nosepiece**. Each objective lens gives a different magnification (for example, 4×, 10×, 40×). By switching lenses, we can zoom in closer or farther.
- **Stage**: This is the flat platform where we place the microscope slide. The stage often has clips to hold the slide in place. These are called **stage clips**.
- **Light source/mirror**: Provides the light needed to pass through the slide so we can see the specimen clearly.
- **Diaphragm**: Adjusts the amount of light that reaches the specimen, helping us see details better.
- **Coarse focus knob**: Moves the stage up and down quickly to bring the specimen into rough focus. It is bigger than the fine focus knob.
- **Fine focus knob**: Moves the stage more slowly to sharpen the image and bring details into clear view.
- **Arm and base**: These support the microscope and keep it steady. The arm is also where we hold the microscope when carrying it.

ADD SOME COLOR! Color code each label and its part of the microscope

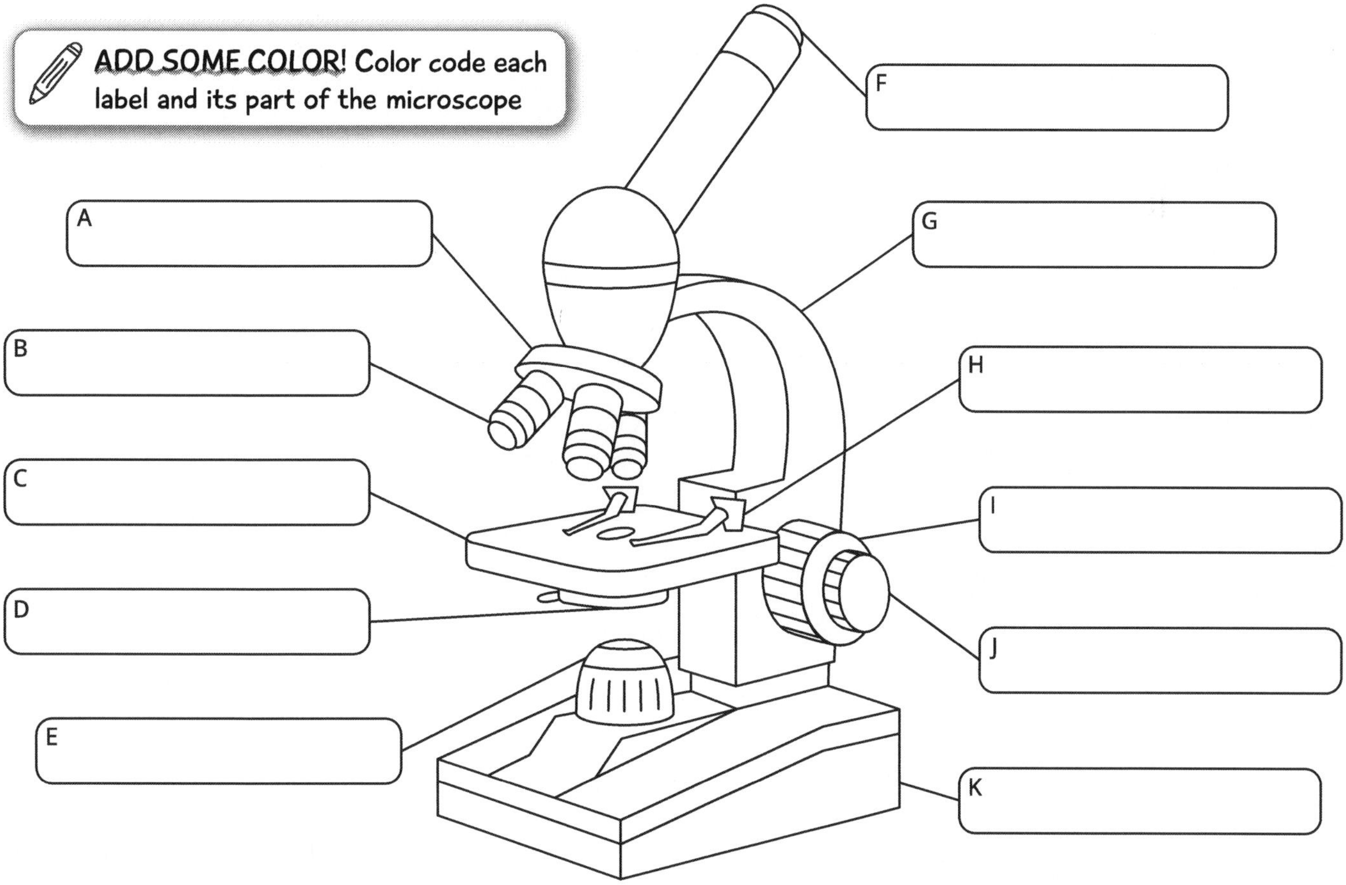

© Emma The Teachie Limited 2025-Present. All rights reserved.

EXTEND → **Let's go deeper on this topic and extend your thinking — like a scientist!**

In medieval times, it was widely believed that some life forms could spontaneously generate — e.g., fleas from dust and maggots from dead flesh. It wasn't until the 1850's that it was disproved. Read about Louis Pasteur's experiment that disproved it, then answer the questions below.

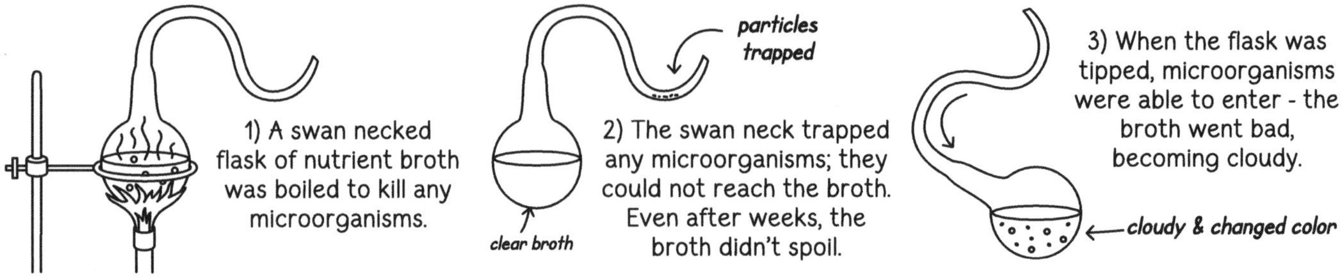

1. The broth was boiled to kill any microorganisms, like bacteria. What is this process called?

____

2. Explain how this experiment disproves the idea of spontaneous generation.

____

____

____

3. Design your own experiment to find evidence that maggots cannot spontaneously arise from dead flesh, using this equipment: glass jars, a lid, mesh, elastic band, 3 pieces of meat, and flies. Include labeled diagrams to show your set up, and then draw or explain the expected results.

NEED TO KNOW: Flies lay eggs, which hatch into maggots in around 24hrs, depending on the species!

REVIEW → **Check & strengthen your understanding by working through this activity.**

Complete these sentences by filling in the missing words.

All cells come from ____________________ cells.

The cell is the basic unit of ____________________ in all organisms.

All organisms are composed of one or more ____________________.

 © Emma The Teachie Limited 2025-Present. All rights reserved.

QUIZ → **Show what you know by answering each of the questions.**

Put your noggin to the test!

1 Decide if each of these statements is TRUE or FALSE.

"Most cells are too small to be seen by the naked eye."

A

"New life can spontaneously arise from nonliving matter."

B

"Electron microscopes are the type of microscope used to view cells in school classrooms."

C

"All organisms are composed of one cell."

D

2 Which of these is NOT a component of cell theory?

A. All cells come from preexisting cells

B. The cell is the basic unit of structure and organization in living organisms

C. All organisms are made of one or more cells

D. All cells are capable of dividing to produce new cells

3 Which scientist concluded that all plants are made of cells?

A. Virchow

B. Schleiden

C. Janssen

D. Hooke

4 Which scientist came up with the term "cells"?

A. Leeuwenhoek

B. Hooke

C. Virchow

D. Schwann

5 Why was the microscope important to the development of cell theory?

____

____

6 What is the flat platform called where the slide is placed?

____

7 Which two parts of the microscope are used to focus the image of the specimen?

____

____

© Emma The Teachie Limited 2025-Present. All rights reserved.

Eukaryotic & Prokaryotic Cells

PROCESS → **Read the text passage below, then complete these tasks to process the info.**

- HIGHLIGHT NEW KEYWORDS
- SQUIGGLY UNDERLINE OLD KEYWORDS
- UNDERLINE ANY DEFINITIONS
- ADD ANY RELEVANT DOODLES IN SPACES

All cells are either eukaryotic or prokaryotic. The defining feature is whether or not the cell has a nucleus.

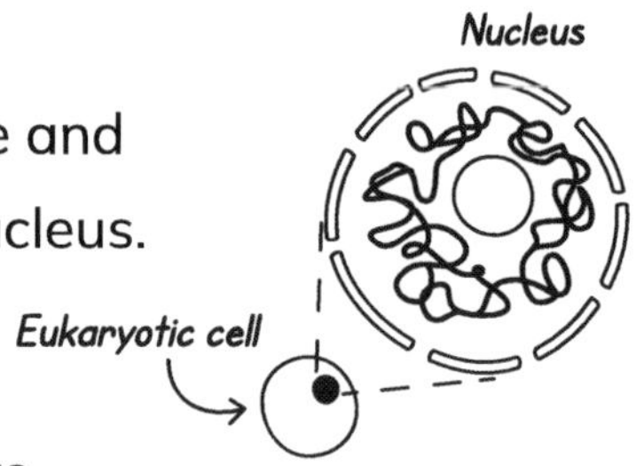

Eukaryotic comes from the Greek words "eu" meaning true and "karyon" meaning kernel — in this case, referring to the nucleus. So, eukaryotic cells are cells that have a nucleus.

Prokaryotic cells do not have a nucleus. "Pro" means before, which refers to the fact that prokaryotes evolved before eukaryotes, and this is why they do not have a nucleus (this came later).

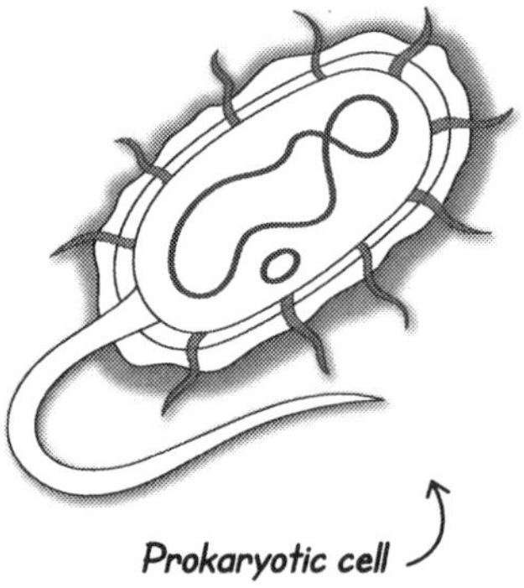

Size

Eukaryotic cells are larger than prokaryotic cells. They are around 10 µm–100 µm, while prokaryotic cells are mostly around 0.1 µm–5.0 µm. In cell volume, eukaryotic cells are often 1000 times bigger than prokaryotic cells.

Unicellular vs. Multicellular

Eukaryotic cells can form multicellular organisms, like animals, plants, and most fungi. However, there are some unicellular eukaryotes — most belong to the group called protists, which includes amoeba and algae. Prokaryotic cells are always unicellular. Bacteria and Archaea are prokaryotic.

Outer Layers of the Cell

Prokaryotic cells always have a cell membrane surrounded by a cell wall.

© Emma The Teachie Limited 2025-Present. All rights reserved.

Eukaryotic cells always have a cell membrane, but don't always have a cell wall. Plants and fungi are eukaryotic, and they DO have a cell wall.

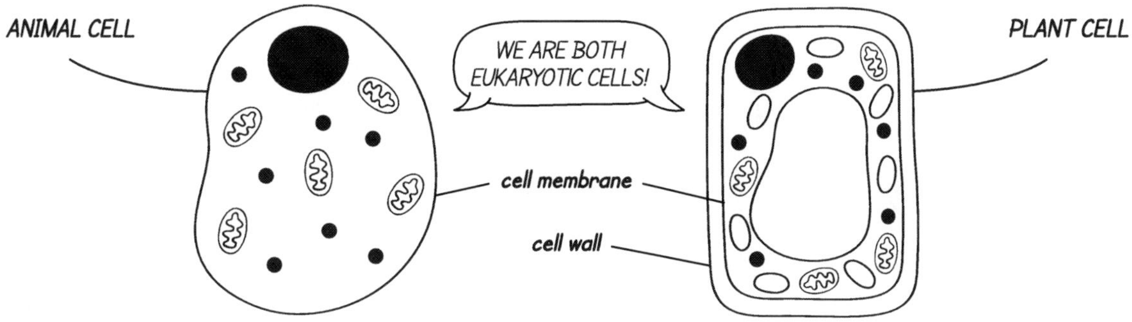

Inside the Cell

Eukaryotic cells have a cytoplasm, and organelles suspended in it. You'll learn about organelles in the next lesson, but these are tiny structures found inside cells. Eukaryotic cells can contain mitochondria, ribosomes, and chloroplasts (in plants and some other eukaryotic cells).

Prokaryotic cells also have a cytoplasm and ribosomes, but they do not have any mitochondria or chloroplasts.

The DNA in eukaryotic cells is found in a nucleus. In prokaryotic cells, there is no nucleus. Their DNA is a single molecule, found free in the cytoplasm. They often have additional rings of DNA, called plasmids.

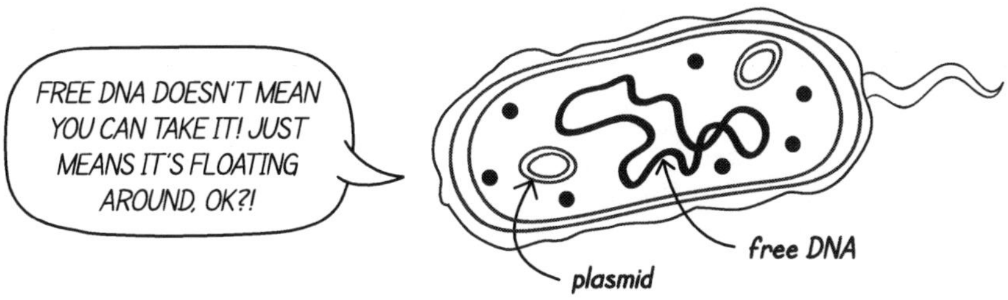

Types of Cell Division

Cells divide to produce more cells. Eukaryotic cells use mitosis to divide, while prokaryotic cells use binary fission. You'll learn more about cell division later!

© Emma The Teachie Limited 2025-Present. All rights reserved.

ACTIVATE → **Complete the activities to apply what you've learned from the text passage.**

1 Add labels to this diagram of a prokaryotic cell. Be sure to check your spelling of any new keywords!

Use a ruler for your arrows, and make sure the arrow touches the part you are labeling!

TAKE IT FURTHER: Some prokaryotes also have a sticky capsule as their outermost layer. They can have pili — small rod-like extensions that help attach to surfaces. And they can have a tail-like structure, called a flagellum, that helps it move! Add all three of these to the diagram, with labels.

2 These mean cells are wanted for their "cellfish" acts of crime! Help the town sheriff write their wanted posters to catch them! Include the name of each cell & its defining features.

© Emma The Teachie Limited 2025-Present. All rights reserved.

3

Read each statement and decide if it if applies to eukaryotes, prokaryotes, or both. Put the number of the statement in the right part of the Venn diagram. If it applies to both eukaryotes and prokaryotes, it goes in the middle overlapping section.

1. Contains a nucleus.
2. Usually between 0.1 μm–5.0 μm in size.
3. Has a cell membrane, a cytoplasm, and ribosomes.
4. All multicellular organisms belong to this group.
5. Can be unicellular.
6. Always has a cell wall.
7. Has circular DNA that is unbound (no nucleus).
8. Contains mitochondria.
9. Bacteria and archaea belong to this group.
10. Use mitosis to divide to make new cells.
11. Contains genetic material.
12. Plants, fungi, and protists belong to this group.

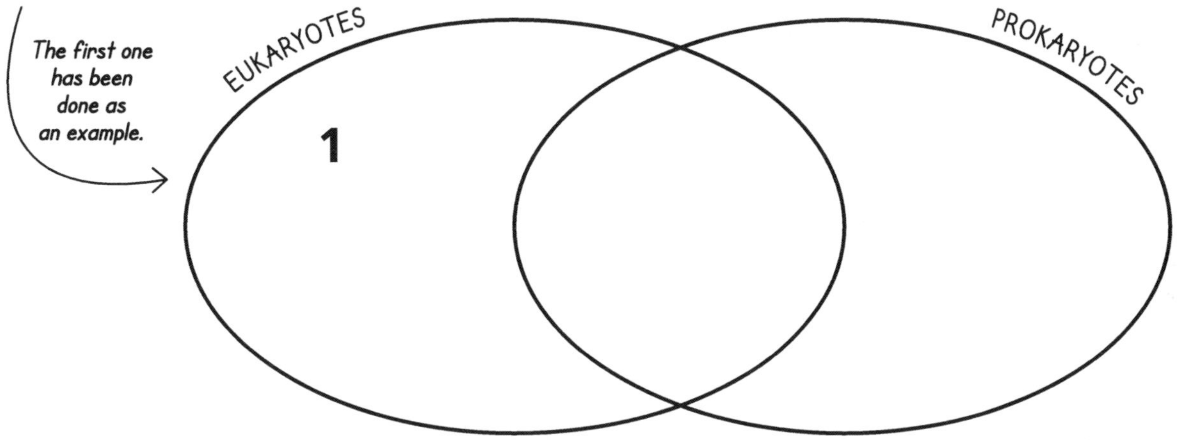

4

Decide if each of these organisms is eukaryotic or prokaryotic. Look closely! Add your own organisms for G. and H., stating if they are eukaryotic or prokaryotic.

A.

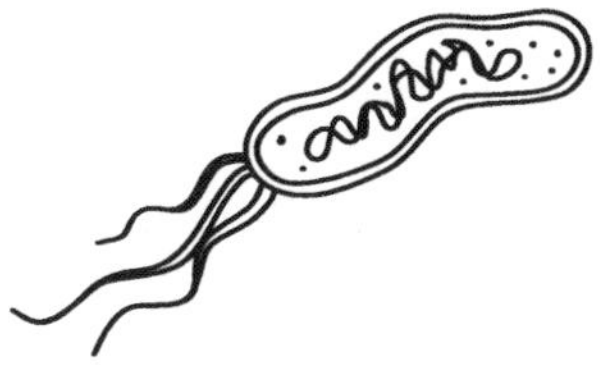

B.

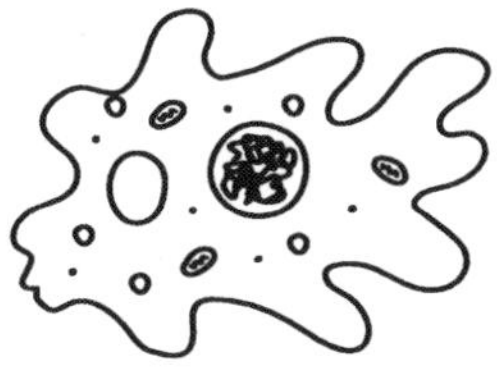

C.

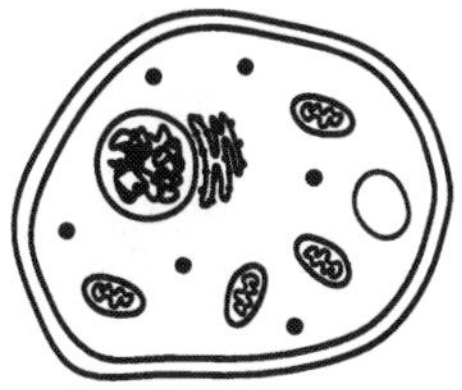

D.

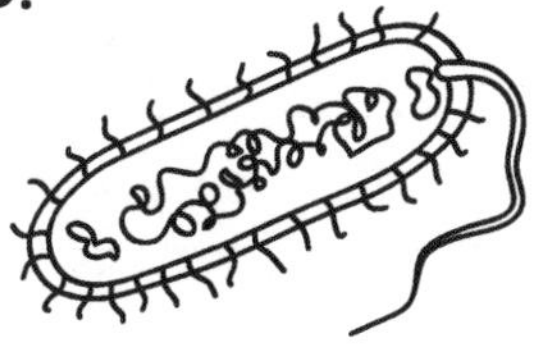

E.

F.

G.

H.

© Emma The Teachie Limited 2025-Present. All rights reserved.

EXTEND → **Let's go deeper on this topic and extend your thinking — like a scientist!**

As mentioned in the activities, some prokaryotic cells have a capsule — this is another layer, made of carbohydrates, that is on top of the cell wall. It can be quite slimey!

List as many possible functions (jobs) of the capsule as you can come up with. No idea is a bad one!

Then, do a few minutes of research. Look up "functions of prokaryotic capsule" and add a check mark to any of your ideas that turned out to be true. Write down any other functions you learn!

REVIEW → **Check & strengthen your understanding by working through this activity.**

Compare eukaryotic and prokaryotic cells by completing the table.

FEATURE	EUKARYOTIC CELL	PROKARYOTIC CELL
1. Has a nucleus?		
2. Size (μm) range		
3. Can be multicellular?		
4. DNA location		
5. Membrane bound organelles present? (e.g., mitochondria, chloroplasts)		
6. Cell wall always present?		
7. Cell division type		

© Emma The Teachie Limited 2025-Present. All rights reserved.

QUIZ → **Show what you know by answering each of the questions.**

1 Identify each structure that the arrow points to.

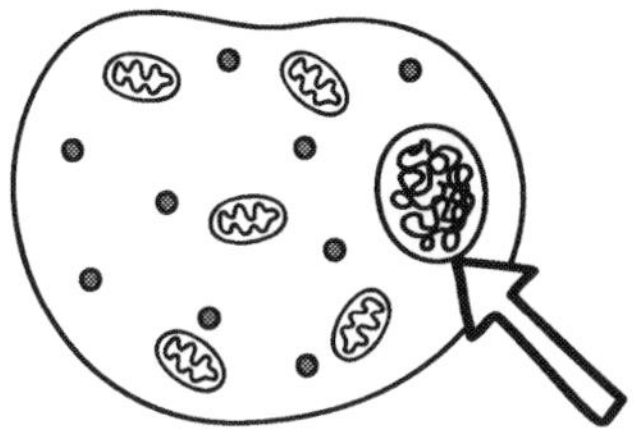

A

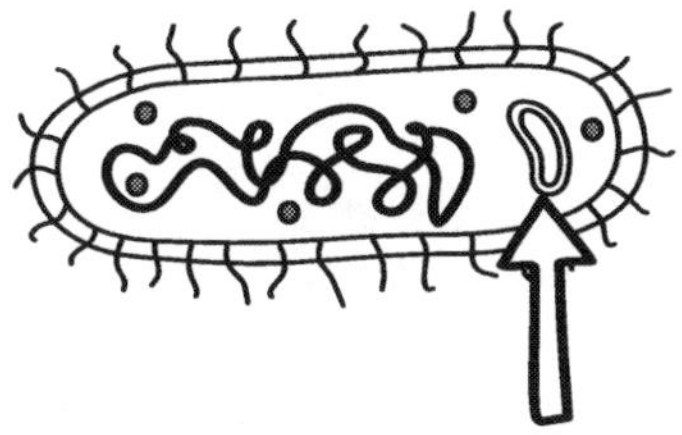

B

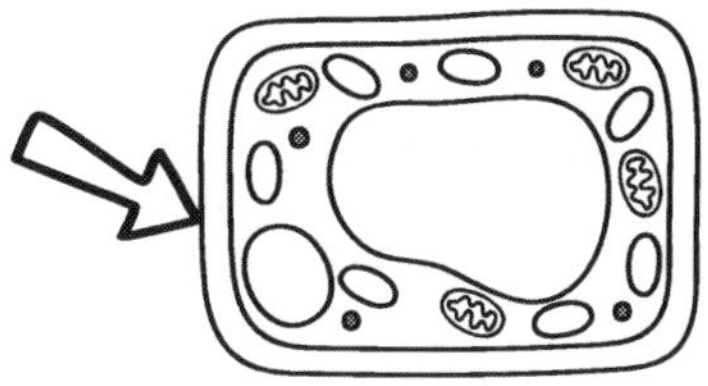

C

D

2 What is the defining feature of a eukaryotic cell?

A. It has a cell membrane

B. It has mitochondria

C. It has a cell wall

D. It has a nucleus

3 Which group of organisms are all prokaryotes?

A. Animals and fungi

B. Amoeba and algae

C. Bacteria and archaea

D. Plants and protists

4 Where is DNA found in a prokaryotic cell?

A. In the nucleus

B. Free in the cytoplasm

C. In mitochondria

D. In chloroplasts

5 Define "eukaryotic" and "prokaryotic."

____

____

____

6 Which process do prokaryotes use for cell division?

____

7 Which type of cell is usually much larger: eukaryotic or prokaryotic?

____

© Emma The Teachie Limited 2025-Present. All rights reserved.

Cell Organelles

PROCESS → **Read the text passage below, then complete these tasks to process the info.**

HIGHLIGHT NEW KEYWORDS — SQUIGGLY UNDERLINE OLD KEYWORDS

UNDERLINE ANY DEFINITIONS — ADD ANY RELEVANT DOODLES IN SPACES

Cell organelles are the tiny structures that we see inside cells when we zoom in with a microscope. They each have a special role in the cell. We'll also look at the cell membrane and the cytoplasm — these aren't technically organelles, but every cell that exists has them!

The Cell Membrane

The cell membrane is the layer that holds everything inside the cell. It separates the inside of the cell from its outside environment. The cell membrane is very thin and flexible. It is made up of a double layer of phospholipids. It controls what enters and exits the cell.

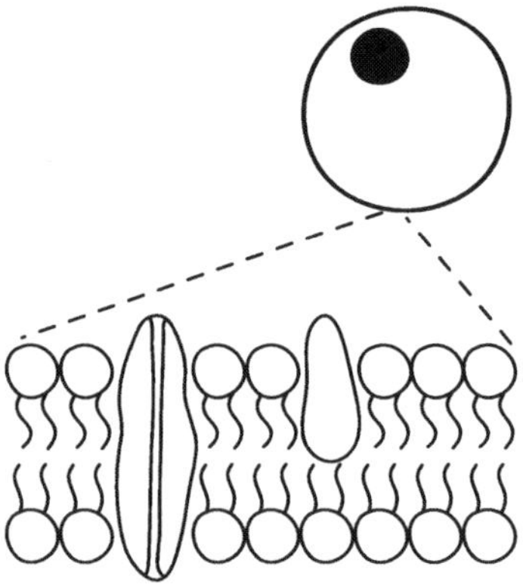

The cell membrane

The Nucleus

The nucleus is sometimes referred to as the "brain" of the cell. It is where the DNA of eukaryotic cells is found. It controls the cell's activities. The nucleus has a double membrane, which is called a nuclear envelope. This has tiny holes in it, called nuclear pores.

> TOP TIP: In simple diagrams, the nucleus is often represented as a black blob! This is because when we use stain to view cells under a microscope, the nucleus stains darker due to the DNA and proteins in it.

The Cytoplasm

The cytoplasm is the fluid that fills the cell. It is mainly made of water, and contains dissolved ions and salts. Organelles are suspended in the cytoplasm. The cytoplasm is important for transporting substances around the cell.

© Emma The Teachie Limited 2025-Present. All rights reserved.

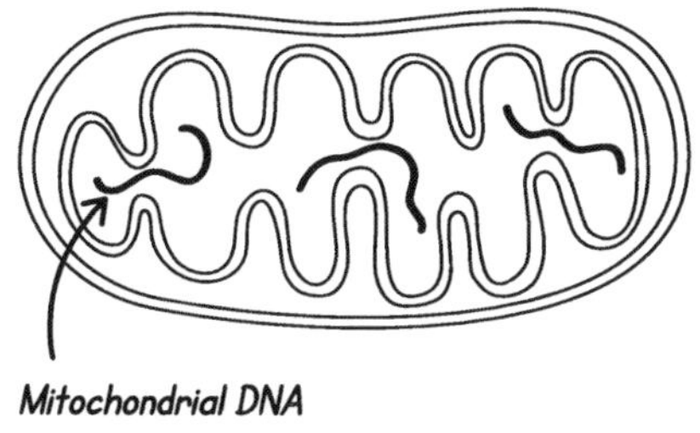

Mitochondrion

There are many mitochondria found in eukaryotic cells. These are the tiny powerhouses of the cell. This means they produce energy for the cell's activities. They do this by carrying out cellular respiration. Mitochondria have their own DNA and a double membrane.

TOP TIP: mitochondrion = singular (one); mitochondria = plural (multiple)

Ribosome

Ribosomes are the protein factories of cells. They can be found attached to the rough endoplasmic reticulum, or they can be found freely suspended in the cytoplasm. Free ribosomes make proteins for the cell's own use, while ribosomes bound to rough endoplasmic reticulum make proteins for transport out of the cell.

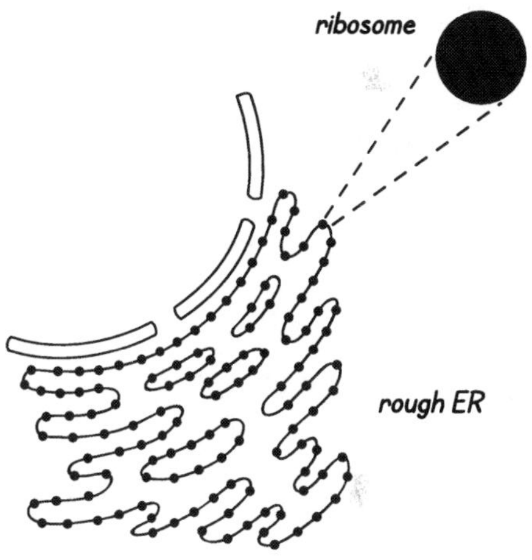

Rough Endoplasmic Reticulum

The rough endoplasmic reticulum (which can also be shortened to "rough ER" or "RER") is also found in animal and plant cells. It is a series of flattened sacs, studded with ribosomes. The rough ER is usually found next to the nucleus, where it is involved in making proteins.

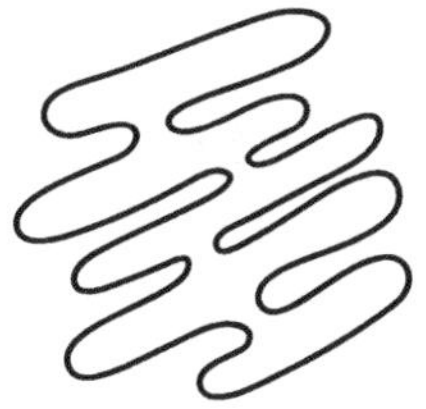

Smooth Endoplasmic Reticulum

The smooth endoplasmic reticulum, also known as "smooth ER" or "SER," is a series of interconnected, flattened sacs. It is found in both plant and animal cells. It is called smooth because it does not have any ribosomes. The smooth ER produces carbohydrates, lipids, and some hormones. It also detoxifies drugs and poisons.

© Emma The Teachie Limited 2025-Present. All rights reserved.

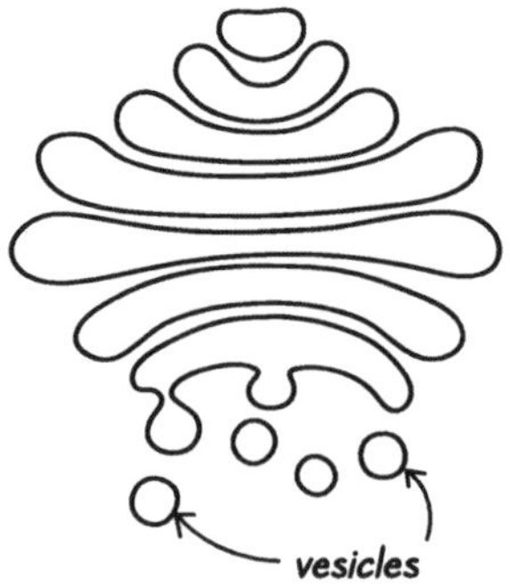

Golgi Apparatus / Golgi Body

The Golgi apparatus is found near the endoplasmic reticulum. It consists of a series of flattened sacs, stacked next to one another. The Golgi apparatus receives vesicles from the ER. These vesicles have proteins and lipids inside. The vesicles fuse with the Golgi apparatus and empty their contents into the sacs. As the proteins and lipids move through the Golgi apparatus, they are modified, tagged, and packaged. They bud off as vesicles on the other side and are transported to other parts of the cell or to the cell membrane for release.

Lysosome

Materials found within cells need to be recycled. In animal cells, lysosomes are the recycling centers of the cell. Plant cells do not have lysosomes. Lysosomes are small vesicles surrounded by a single membrane that contain digestive enzymes. These are used in the breakdown of worn-out organelles. They also help to digest whole cells and foreign particles that enter the cell.

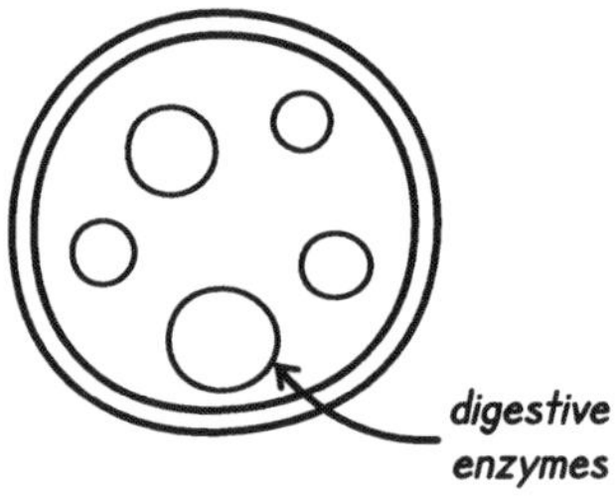

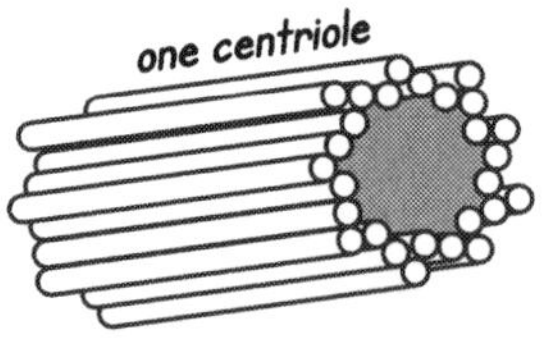

Centrioles / Centrosome

Centrioles are found only in animal cells. They are bundles of microtubules arranged in a cylinder shape. Two centrioles make up one centrosome. Centrosomes play an important role in cell division. They are responsible for forming and organizing spindle fibers.

Vacuole

Vacuoles are membrane-bound sacs. Both animal and plant cells can have vacuoles. Plant cells have one large permanent vacuole, whereas animal cells may have lots of smaller temporary vacuoles. In animal cells, the function of vacuoles is

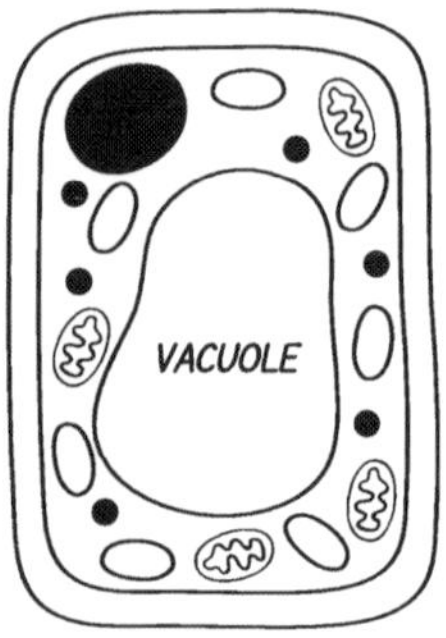

© Emma The Teachie Limited 2025-Present. All rights reserved.

mainly to store waste products, like salts. In plant cells, the large permanent vacuole has an important role in providing support to the cell, keeping it turgid. It also stores water, cell sap, and waste products. Like the lysosome in animal cells, the vacuoles in plant cells act as recycling centers.

Chloroplast

Chloroplasts are an organelle only found in plant and algal cells. Not all plant cells contain chloroplasts, e.g., root hair cells. Chloroplasts are where photosynthesis takes place. They contain a green pigment called chlorophyll. Chlorophyll absorbs light for photosynthesis to occur, making glucose (food) for the plant.

In textbooks, chloroplasts are shown in cells as small green oval shapes.

Cell Wall

Animal cells do not have a cell wall. Plant cells always have a cell wall. This is the outermost layer of the plant cell, found outside the cell membrane.

TOP TIP: To remember that the cell wall is outside of the cell membrane, think of the wall of your house; it is always the last layer on the outside!

In plants, the cell wall is made of a fiber called cellulose. In fungi, the cell walls are made of chitin. The main function of the cell wall is to provide structural strength to the cell. It prevents the cell from bursting when an abundance of water enters through osmosis. The cell wall can have small holes. These holes connect the cytoplasm of neighboring cells, allowing them to communicate and transport substances.

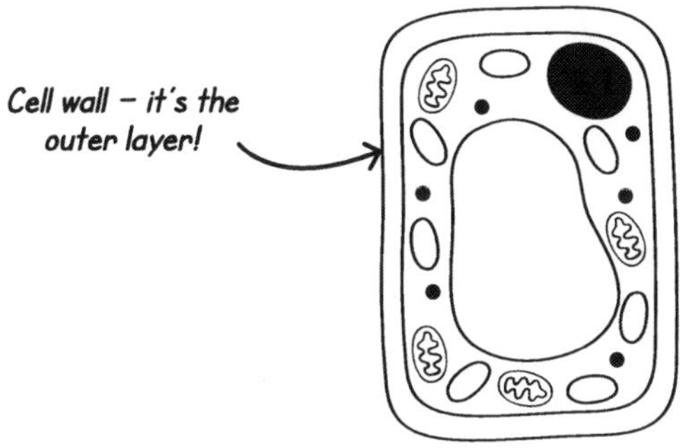

© Emma The Teachie Limited 2025-Present. All rights reserved.

ACTIVATE ➔ **Complete the activities to apply what you've learned from the text passage.**

1 Label these diagrams of an animal and a plant cell. Then, add a star to any organelles or structures that are only found in the plant cell and not in the animal cell.

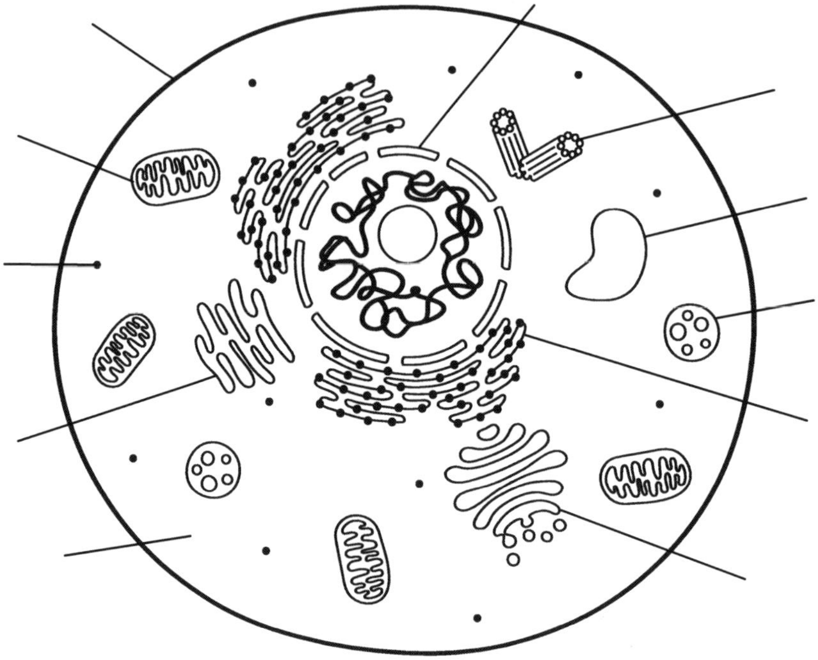

We use simplified diagrams of organelles – there is more detail than this, but it would be hard to include all of it in one picture!

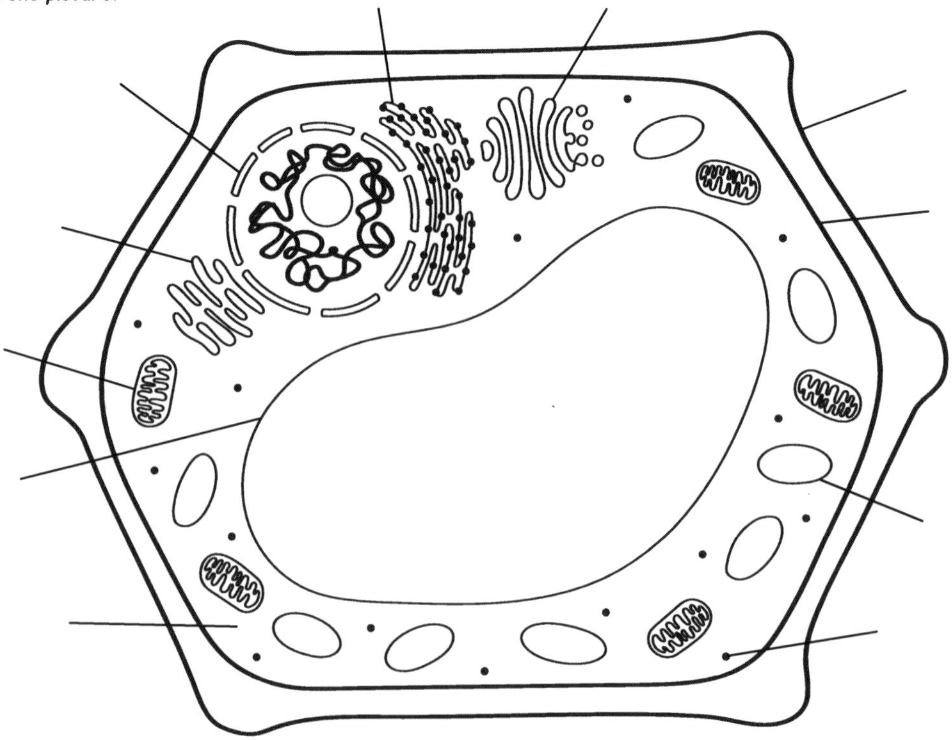

© Emma The Teachie Limited 2025-Present. All rights reserved.

2

It's really important to understand what job each organelle does in the cell. Describe the function of each organelle/cell structure using all of the words in the word box.

powerhouse	respiration	photosynthesis	packages	waste		
stores	brain	DNA	controls	proteins	strength	division

1. Cell membrane: ______________________________
2. Cell wall: ______________________________
3. Nucleus: ______________________________
4. Ribosome: ______________________________
5. Smooth ER: ______________________________
6. Rough ER: ______________________________
7. Mitochondrion: ______________________________
8. Golgi apparatus: ______________________________
9. Lysosome: ______________________________
10. Vacuole: ______________________________
11. Chloroplast: ______________________________
12. Cytoplasm: ______________________________
13. Centrioles: ______________________________

3

Let's compare the components of animal cells and plant cells. Place a check mark in each row to show if the structure is found in plant cells, animal cells, or both.

Structure	Animal cell	Plant cell	Both
1. Nucleus	☐	☐	☐
2. Cell wall	☐	☐	☐
3. Mitochondria	☐	☐	☐
4. Cell membrane	☐	☐	☐
5. Chloroplasts	☐	☐	☐
6. Lysosome	☐	☐	☐
7. Large central vacuole	☐	☐	☐
8. Centrioles	☐	☐	☐

© Emma The Teachie Limited 2025-Present. All rights reserved.

EXTEND ➡ **Let's go deeper on this topic and extend your thinking — like a scientist!**

The image on the right is of a unicellular eukaryotic organism called "Euglena." Look closely at what it contains.

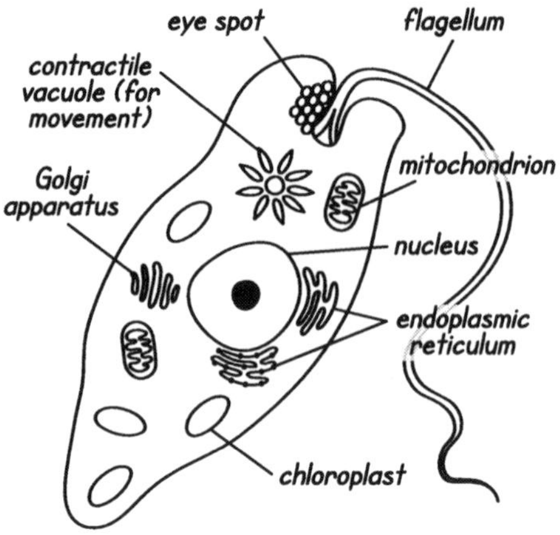

1. Which components does Euglena contain that suggest it is a plant cell?

2. Which components does Euglena contain that suggest it is an animal cell?

3. As Euglena has features of both animal cells and plant cells, it cannot be classified as either. Which other type of eukaryotic organism could it be?

REVIEW ➡ **Check & strengthen your understanding by working through this activity.**

Decide if each statement is TRUE or FALSE — circle T or F. If you think it is false, explain why.

STATEMENT	TRUE OR FALSE	If you circled false, why is it false?
1. Ribosomes can only be found attached to the rough ER.	T / F	
2. Plant and animal cells both contain mitochondria.	T / F	
3. The cytoplasm is for producing proteins.	T / F	
4. The cell membrane is rigid and made of cellulose.	T / F	
5. Lysosomes recycle materials in animal cells.	T / F	
6. Chloroplasts are where photosynthesis happens.	T / F	
7. The nucleus controls what enters and exits the cell.	T / F	
8. Only plant cells have vacuoles.	T / F	

© Emma The Teachie Limited 2025-Present. All rights reserved.

QUIZ → **Show what you know by answering each of the questions.**

Put your noggin to the test!

1 Identify each of the organelles shown in the images.

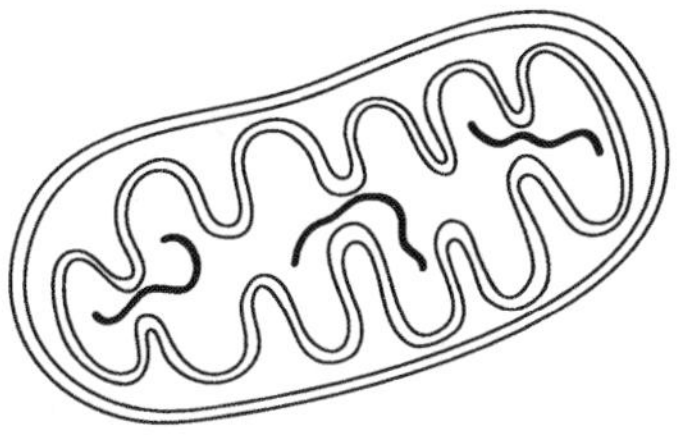

A

B

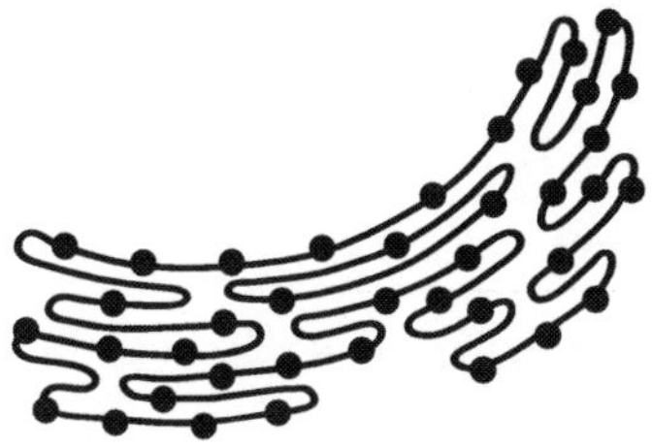

C

D

2 Which organelle is the site of photosynthesis?

A. Rough ER

B. Chloroplast

C. Mitochondrion

D. Vacuole

3 Which organelle is the powerhouse of the cell?

A. Mitochondrion

B. Golgi apparatus

C. Smooth ER

D. Nucleus

4 Which organelle is only found in animal cells?

A. Lysosome

B. Vacuole

C. Nucleus

D. Chloroplast

5 What is the role of the nucleus?

____

____

6 Why do plant cells need a cell wall?

____

____

7 What is the function of ribosomes?

____

____

© Emma The Teachie Limited 2025-Present. All rights reserved.

Specialized Cells

PROCESS → **Read the text passage below, then complete these tasks to process the info.**

HIGHLIGHT NEW KEYWORDS

SQUIGGLY UNDERLINE OLD KEYWORDS

UNDERLINE ANY DEFINITIONS

ADD ANY RELEVANT DOODLES IN SPACES

Specialized cells are cells in the body that have developed a unique structure to help carry out their particular function. Multicellular organisms have many types of specialized cells, as different cells are needed to carry out different functions. For example, muscle cells are needed for movement, and nerve cells are needed for sending information around the body.

Stems Cells

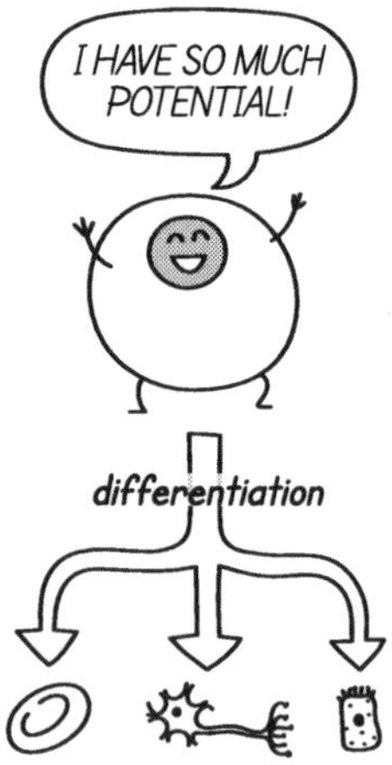

Before cells become specialized, they are undifferentiated — this means they do not have a particular structure yet. We call them stem cells.

Stem cells have the potential to turn into other types of cells. This is known as differentiation. They differentiate by expressing different genes, which changes the structure and makes them ready to carry out a particular function. This happens in an embryo. As the embryo develops, stem cells differentiate into all the different types of body cells, which build up into tissues and organs.

Differentiation can also happen in some adult tissues. Adult stem cells can differentiate to replace worn out or dead cells, like blood cells.

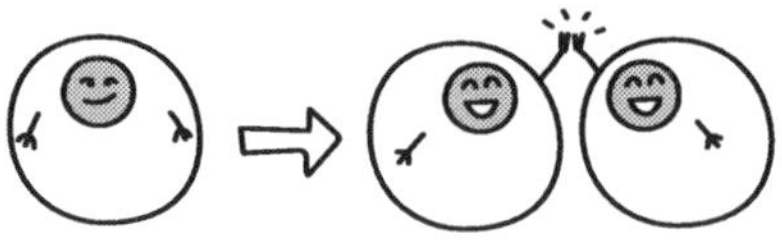

Stem cells can also self-renew, which means they divide to produce more stem cells.

Examples of specialized cells

Nerve cells: Also known as neurons, nerve cells are responsible for carrying information around the body. The information travels as electrical impulses

© Emma The Teachie Limited 2025-Present. All rights reserved.

along nerve cells. They are adapted for this job by having extensions called axons to carry the impulse over long distances. The axon can be myelinated (covered in a fatty layer of insulation) to speed up the transmission of the electrical impulse.

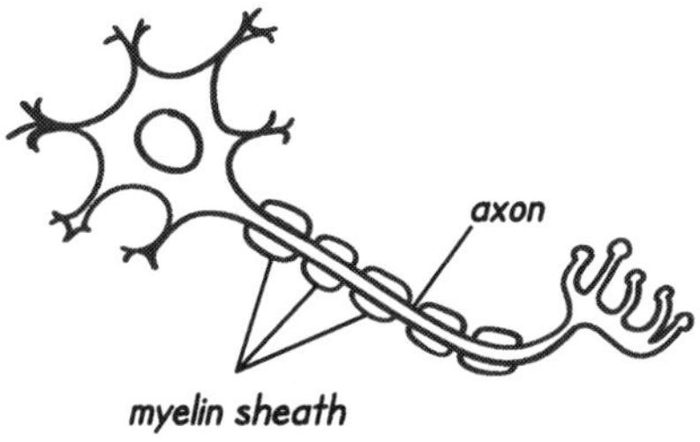

Red blood cells: Also known as erythrocytes, red blood cells are needed to carry oxygen around the body to respiring cells. To do this, they contain a protein called hemoglobin that can bind to oxygen. They also have a concave disc shape that increases their surface area for carrying more oxygen.

Fun fact: Your body makes about 2 million new red cells every second!

Root hair cells: Found in the roots of plants, root hair cells are responsible for absorbing water and mineral ions from the soil. They have a long extension to increase their surface area, which allows them to increase the rate of absorption. And they have lots of mitochondria to release the energy needed to move mineral ions into the cell in the process of active transport (more about this later!).

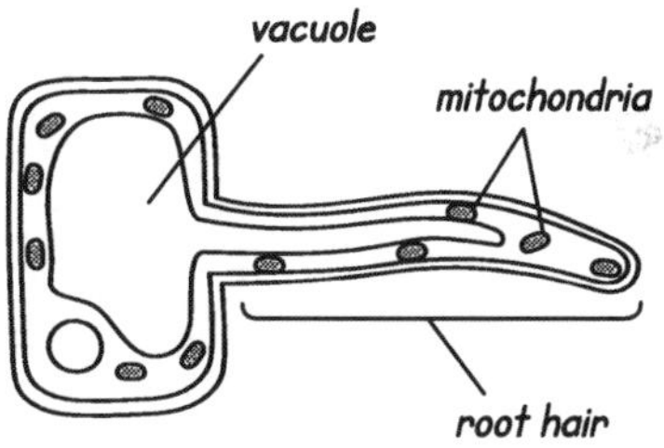

THINK: What is **NOT** present in the root hair cell? Why is it not there?

Muscle cells: Also known as myocytes, muscle cells are needed for movement in animals. They contain special protein filaments that contract and relax to cause movement. They also have many mitochondria to provide the energy needed for the contractions.

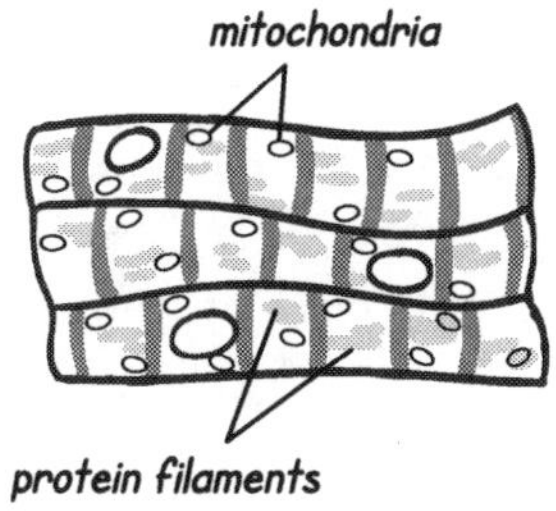

There are many, many more specialized cells in both animals and plants.

TOP TIP: The important skill here is linking structure to function! Ask: What structures and organelles does each type of cell have, and how do these help it carry out its job?

© Emma The Teachie Limited 2025-Present. All rights reserved.

ACTIVATE → **Complete the activities to apply what you've learned from the text passage.**

1 Label this diagram to show the role of stem cells in the body and the two processes they can undergo.

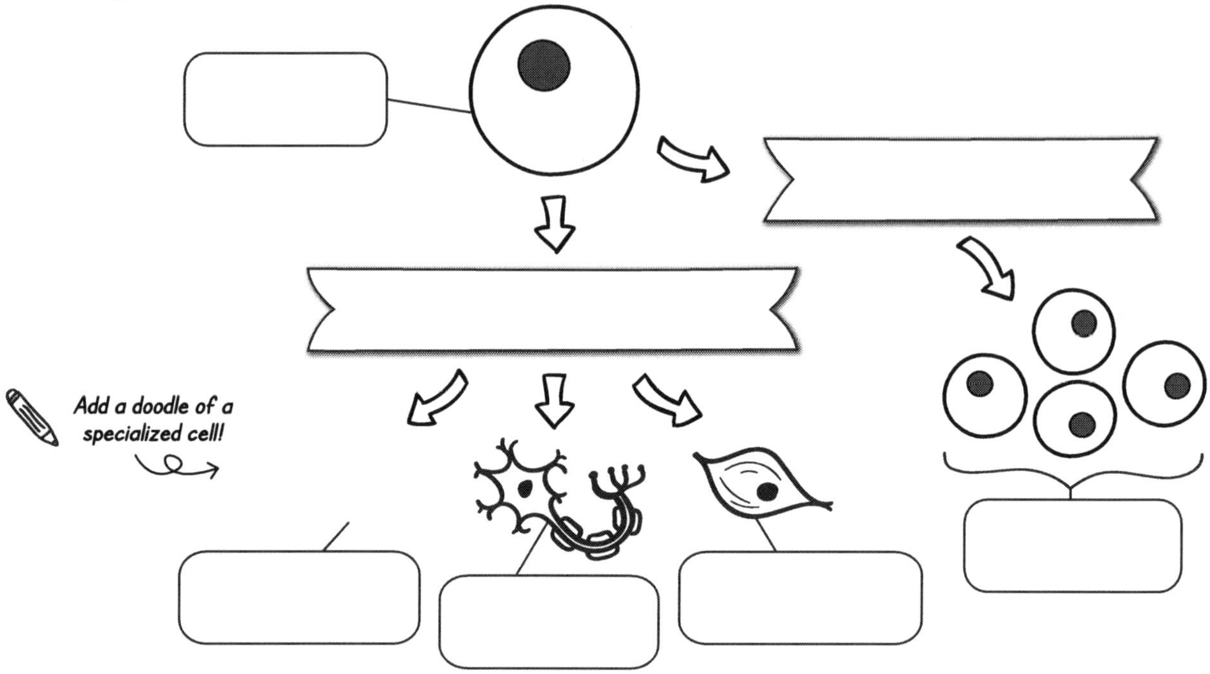

2 It's easy to just think about specialized cells in the human body. But plants have many types of specialized cells, too! Take a close look at the diagram of each type of plant cell, then match it to its structure and its function using arrows.

PICTURE

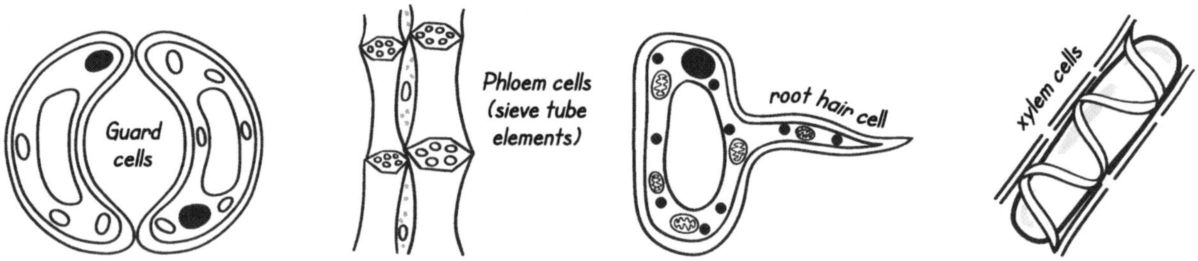

STRUCTURE	No inner organelles to allow sap to flow between them; their end walls have pores (holes) to allow sap to move between cells	Dead hollow cells; their end cell walls break down to form continunous tubes; strengthened by lignin fibers	Bean shaped; found in pairs; have a thicker cell wall on the inner side to allow them to curve when full of water, forming an opening	Long root hair to increase surface area; large vacuole to store water; many mitochondria to provide energy for active transport
FUNCTION	To transport sugars (sap) produced in photosynthesis	To regulate water loss in the leaves by opening and closing stoma (tiny holes in the leaves)	To absorb water and mineral ions from the soil	To transport water from the roots to the rest of the plant

© Emma The Teachie Limited 2025-Present. All rights reserved.

3

Look at the diagram and read the information about sperm cells. Then, explain how the structure of a sperm cell allows it to carry out its function.

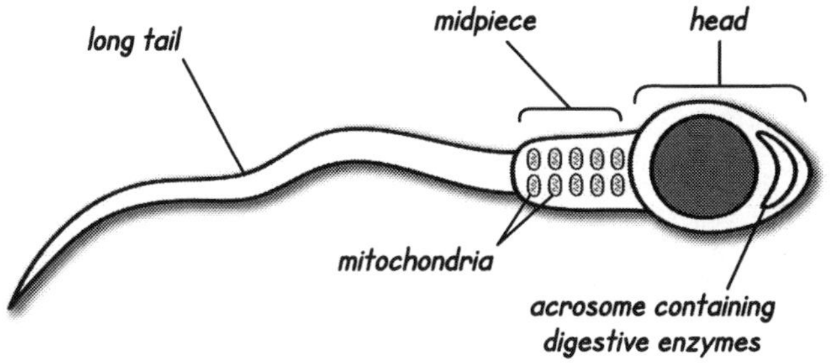

Sperm cells are reproductive cells. Their job is to swim through the female reproductive system to reach and fertilize the egg cell. They have to swim around 1,000-fold their own length.

When they reach the egg cell, they need to penetrate it by breaking down the egg's protective outer layers. Then, the sperm cell's DNA can be combined with the egg cell's DNA.

Explain how the sperm cell is adapted for its function.

____

____

____

____

____

____

____

4

Fill in the missing words to show how each type of specialized cell is adapted for its function.

- Nerve cells have long ____________ to carry electrical impulses.
- Red blood cells contain ____________ to bind oxygen. They have a concave disc shape to increase their ____________ ____________ so more oxygen can bind.
- Root hair cells have many ____________ to provide energy for active transport of mineral ions into the cell. They have a large central ____________ for storing water and ions.
- Muscle cells contain protein ____________ that contract and relax to cause movement. They have many mitochondria to provide the ____________ for this movement.

© Emma The Teachie Limited 2025-Present. All rights reserved.

EXTEND → **Let's go deeper on this topic and extend your thinking — like a scientist!**

Stem cells can be used to help treat medical conditions and disease. Read about bone marrow transplants, then answer the questions below.

Did you know that your blood cells are made in the bone marrow? That's because there are adult stem cells in the bone marrow. When a person has blood cancer, such as leukemia or lymphoma, they may be offered a bone marrow transplant. This involves taking healthy stem cells from the donor's bone marrow and using them to replace the bone marrow of the patient.

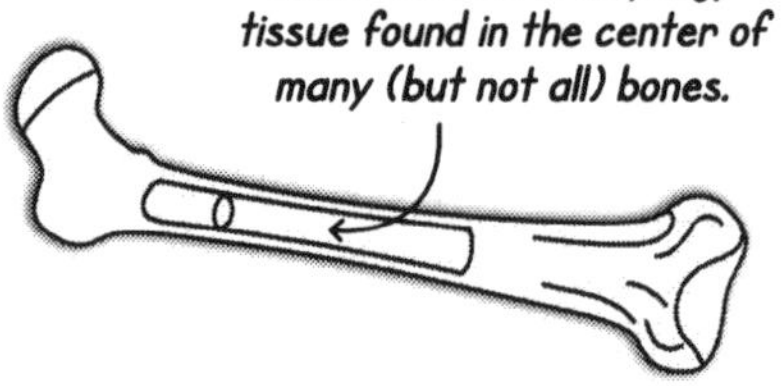

1. Which type of specialized cells are bone marrow stem cells capable of becoming?

____

2. The patient will have treatment to eliminate cancerous cells and unhealthy bone marrow before the donor bone marrow cells are transferred. Using what you know about stem cells, what do you think happens when the patient receives the donor cells?

____

____

____

> TAKE IT FURTHER: Research "peripheral blood stem cells" and find out how these can be collected. Why is this method often used instead of a bone marrow transplant? Are there any drawbacks to using peripheral blood stem cells?

REVIEW → **Check & strengthen your understanding by working through this activity.**

Unscramble the anagram to reveal the correct answer to each statement.

	STATEMENT	ANAGRAM		ANSWER
A.	This cell carries electrical impulses around the body.	RUNNEO	=	________
B.	A cell that has specific features that allow it to carry out a role in the body.	ZAPILDECIES	=	________
C.	This cell is adapted for the absorption of water and mineral ions from the soil.	HELLO ROTI ARC	=	________
D.	Another word for a cell's particular job.	NUTICOFN	=	________
E.	A cell that is capable of differentiating to become a specialized cell.	LEST CLEM	=	________
F.	A cell that has a concave disc shape and high levels of hemoglobin.	DECL ROD LOBEL	=	________

© Emma The Teachie Limited 2025-Present. All rights reserved.

QUIZ → **Show what you know by answering each of the questions.**

1 Identify each type of specialized cell shown in the images.

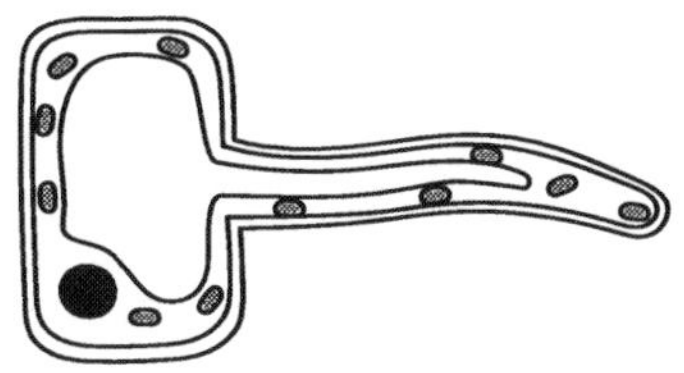

A

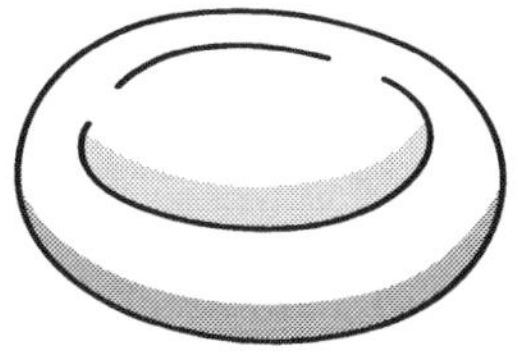

B

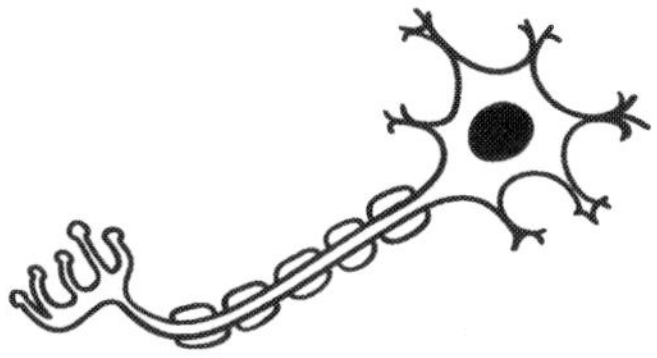

C

D

2 What are stem cells?

A. Cells that are already specialized

B. Cells that can turn into other types of cells

C. Cells that carry impulses

D. Cells only found in the stems of plants

3 Which protein do red blood cells contain?

A. Myelin

B. Hemoglobin

C. Keratin

D. Collagen

4 Which adaptation helps nerve cells carry impulses over long distances?

A. Hemoglobin

B. Filaments

C. Vacuoles

D. Axons

5 Which process describes a stem cell becoming specialized?

6 Why do sperm cells need many mitochondria?

7 How are muscle cells specialized for their function?

© Emma The Teachie Limited 2025-Present. All rights reserved.

UNIT 2 REVIEW CROSSWORD

ACROSS

6 The scientific theory describing the basic principles of cells
9 Jellylike material inside the cell that holds organelles
11 The flat part of a microscope that holds the slide
13 Scientist who first used the word "cell" when viewing cork under a microscope
15 The process of a stem cell changing into a more specific type of cell
18 The brain of the cell that stores DNA and controls cell activities
19 A cell without a nucleus or membrane-bound organelles
20 A cell that has a nucleus and membrane-bound organelles
21 A tiny cell structure that carries out a specific function

DOWN

1 Tiny organelle that makes proteins
2 Outer structure found in plant, fungus, and bacterial cells
3 Cell sac that stores water, food, or waste
4 There is a hierarchy of this in complex organisms
5 Dutch eyeglass makers who invented the first compound microscope
6 Green organelle in plant cells where photosynthesis happens
7 Unspecialized cell that can become many different cell types
8 All cells come from cells that are...
10 Made of only one cell
12 The "powerhouse" of the cell
14 The part of a microscope you look through
16 Having a unique structure and job, like nerve or muscle cells
17 Organelle that breaks down waste and old cell parts

© Emma The Teachie Limited 2025-Present. All rights reserved.

UNIT 3 HOW CELLS WORK

© Emma The Teachie Limited 2025-Present. All rights reserved.

The Cell Membrane

PROCESS ➔ **Read the text passage below, then complete these tasks to process the info.**

HIGHLIGHT NEW KEYWORDS — SQUIGGLY UNDERLINE OLD KEYWORDS

UNDERLINE ANY DEFINITIONS — ADD ANY RELEVANT DOODLES IN SPACES

Our cells need to be able to transport molecules in and out. They also need to be able to communicate with each other and to trigger responses in the body, e.g., to pathogens. The cell membrane handles all of this.

The cell membrane is the gatekeeper of the cell. It decides what can enter and exit the cell. We say that it is "selectively permeable" or semipermeable". Permeable means that something is able to pass through. And selective means that not everything is chosen.

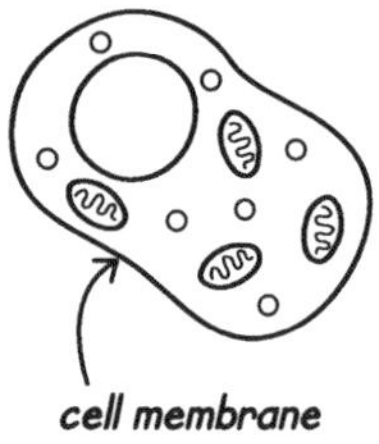

For example: Small, nonpolar molecules, like oxygen and carbon dioxide, can pass freely through the cell membrane.

> Polar = the molecule has charge across it; Nonpolar = no charge across it

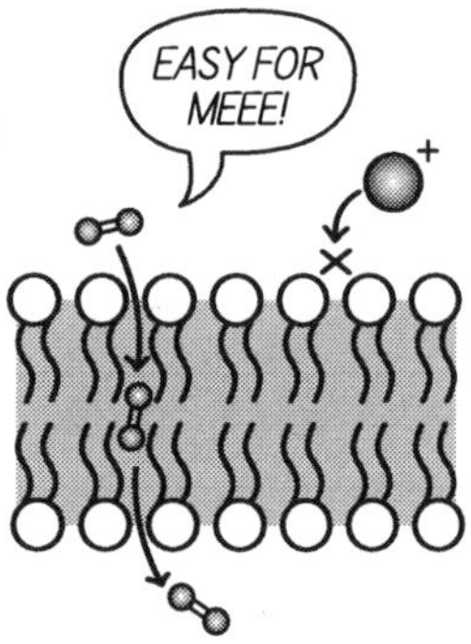

Large or polar molecules, like glucose and proteins, cannot pass freely through the cell membrane. They need to use transport proteins that are embedded in the cell membrane.

By controlling what moves in and out of the cell, the cell membrane plays an important role in homeostasis — maintaining a stable internal environment. This keeps our cells healthy and functioning!

Structure of the Cell Membrane

We use the fluid mosaic model to describe the cell membrane. It is fluid, as the components in it can move around, and it is like a mosaic because it is made of lots of little pieces.

The main component of the cell membrane is phospholipids. These are made

© Emma The Teachie Limited 2025-Present. All rights reserved.

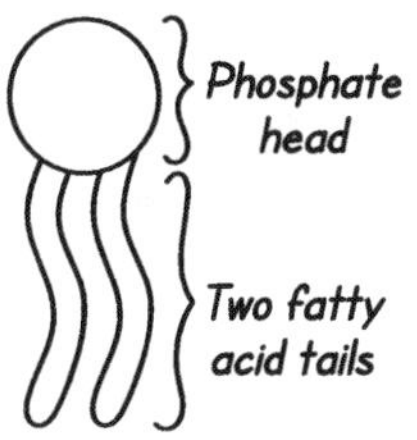

of a hydrophilic phosphate head (hydrophilic = water loving) and two fatty acid tails that are hydrophobic (hydrophobic = water hating). These properties determine what can and cannot pass freely through the cell membrane.

The phospholipids form a double layer, called a bilayer. The fatty acid tails face inwards, away from water, and the phosphate heads face outwards.

There are lots of different molecules embedded in the phospholipid bilayer.

Cholesterol regulates the fluidity of the cell membrane. At low temperatures, it keeps the phospholipids spread out, and at high temperatures, it helps bring them together to stop the cell membrane getting too fluid.

There are transport proteins, like channel proteins and carrier proteins, embedded in the cell membrane — these move large and polar molecules across the cell membrane.

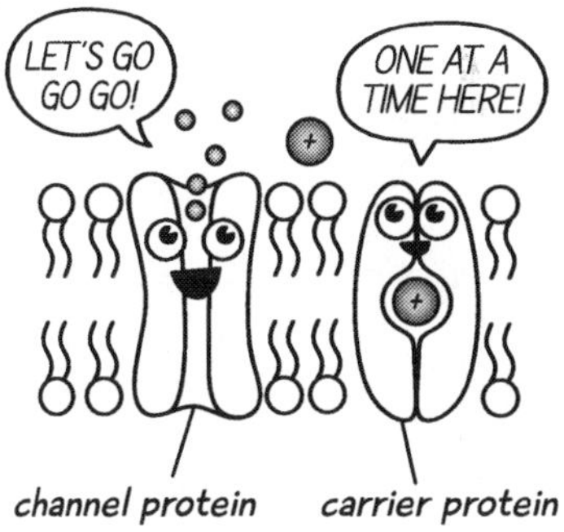

There are also receptor proteins that are responsible for cell signaling. They allow cells to communicate with each other.

The cell membrane plays a role in cell recognition. The immune system seeks to destroy foreign cells and particles. Obviously, we don't want it to destroy our own cells. So various molecules on the surface of the cell membrane act as a marker for the immune system, saying, "Hey, I'm one of you!"

The cell membrane also has molecules that help with cell adhesion — this allows cells to stick to each other to form tissues and organs.

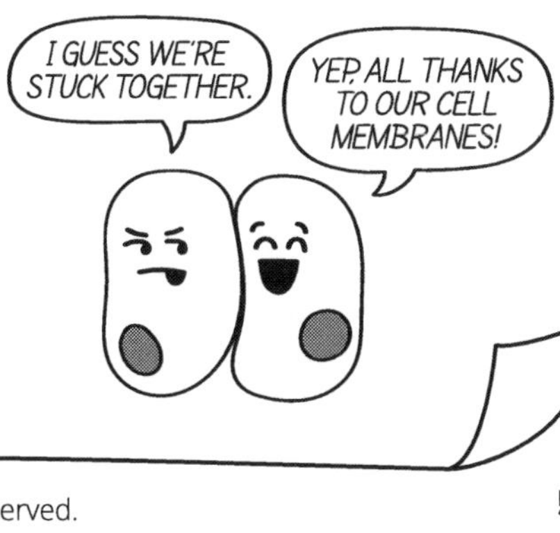

© Emma The Teachie Limited 2025-Present. All rights reserved.

ACTIVATE ➜ **Complete the activities to apply what you've learned from the text passage.**

1 The text passage describes four functions of the cell membrane. List these below and add a doodle or description for each of them.

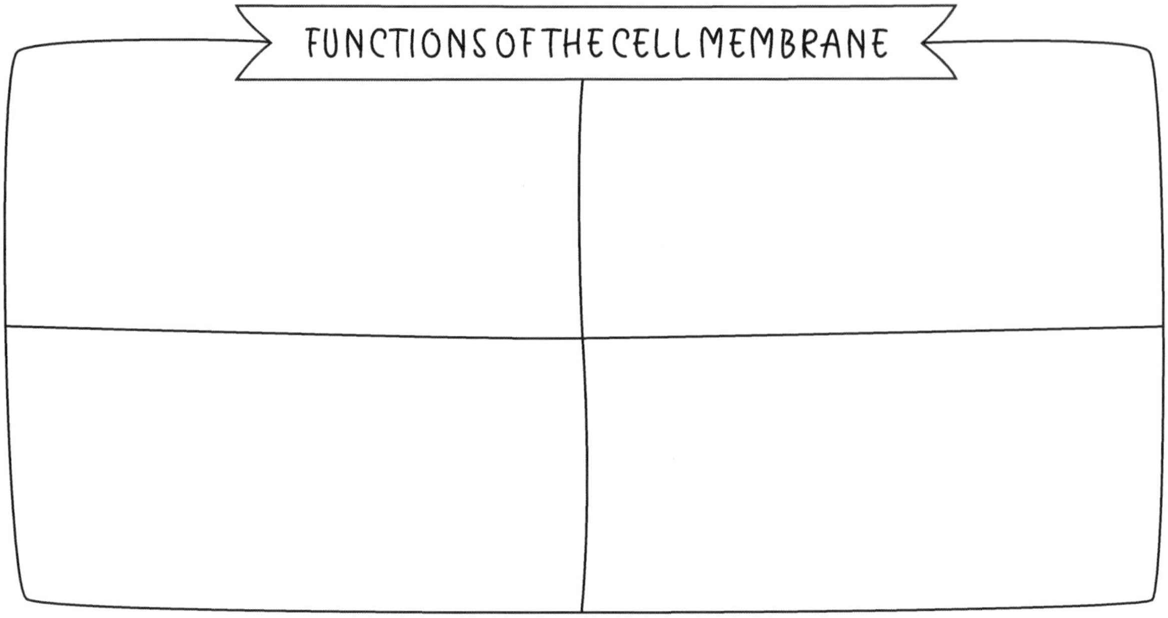

2 Complete the missing labels for this diagram of the cell membrane.

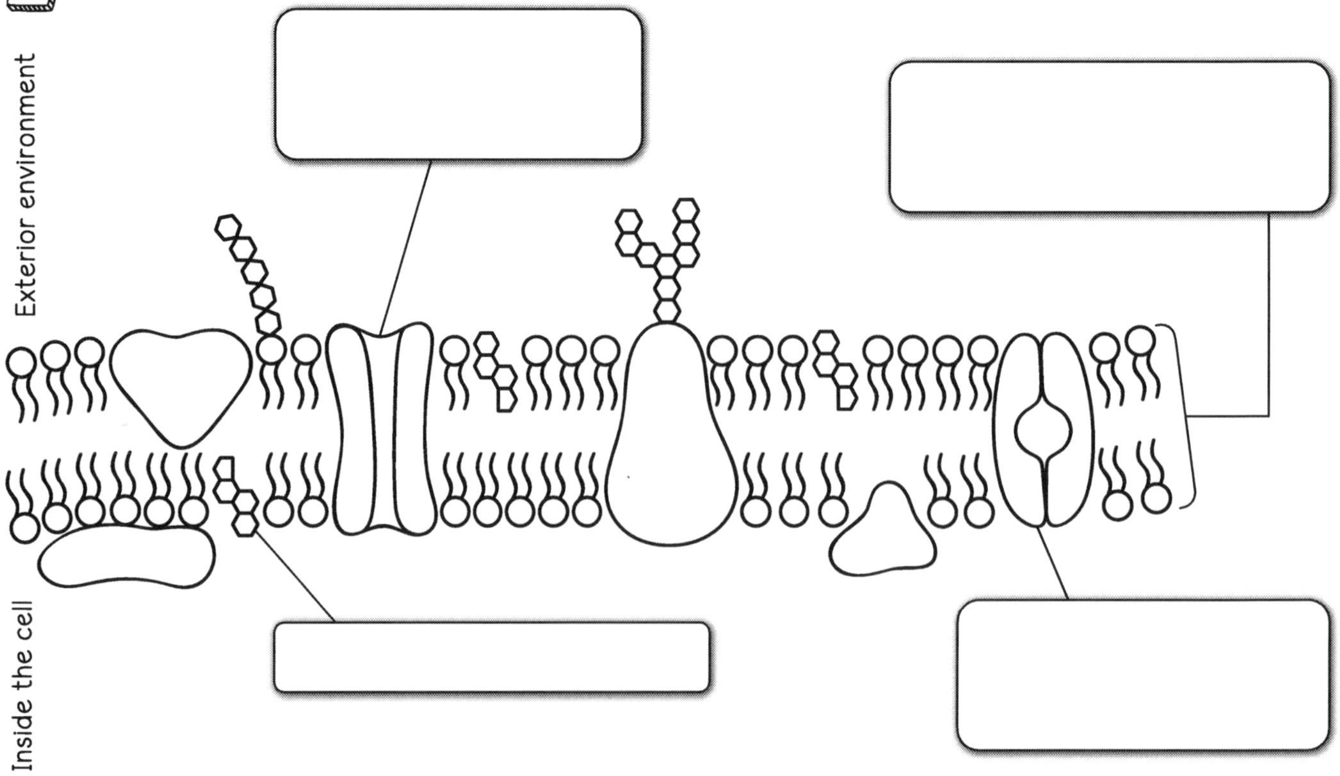

TAKE IT FURTHER: Do extra research and add your own labels for any other components you see!

© Emma The Teachie Limited 2025-Present. All rights reserved.

3 Read the description below each box. Your job is to create a diagram that shows what is being described — i.e., create a storyboard that shows how channel proteins and carrier proteins work. Use the key to help you draw your diagrams (add color!).

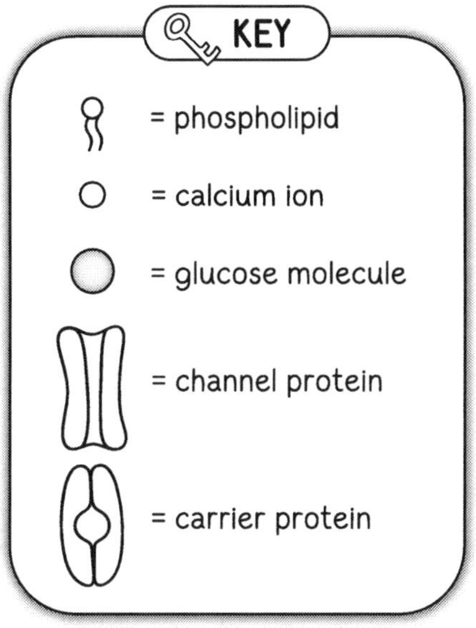

CHANNEL PROTEINS

There are calcium ions (charged molecules) outside the cell. They can move through channel proteins.

The channel protein allows many of the calcium ions to pass through at once and enter the cell.

CARRIER PROTEINS

There are glucose molecules outside the cell. One of them binds to the surface of the carrier protein.

This binding causes the carrier protein to change shape, bringing the glucose through the cell membrane.

The glucose molecule is released inside the cell, and the carrier protein returns to its normal shape.

4 Label the parts of the phospholipid. Then, write speech bubbles for each part of it to demonstrate its properties in relation to water. *(The faces are a hint!)*

Add your speech bubbles here!

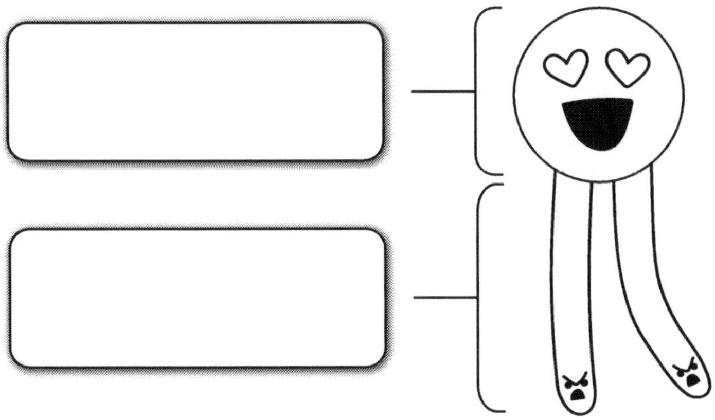

© Emma The Teachie Limited 2025-Present. All rights reserved.

EXTEND → **Let's go deeper on this topic and extend your thinking — like a scientist!**

Read this info about some tricky viruses, then think of solutions — how could you stop this?

The vaccinia virus is able to trick its way into host cells. It does this by displaying molecules on its surface that are usually found on the cell membranes of dead cells. This is very clever, because it allows the viruses to evade the immune system.

Even better, appearing as a dead cell actually encourages host cells to absorb the virus as a way of cleaning up dead cells! Once inside the host cell, the virus can use the cell's machinery to replicate its viral DNA and make components for new viruses.

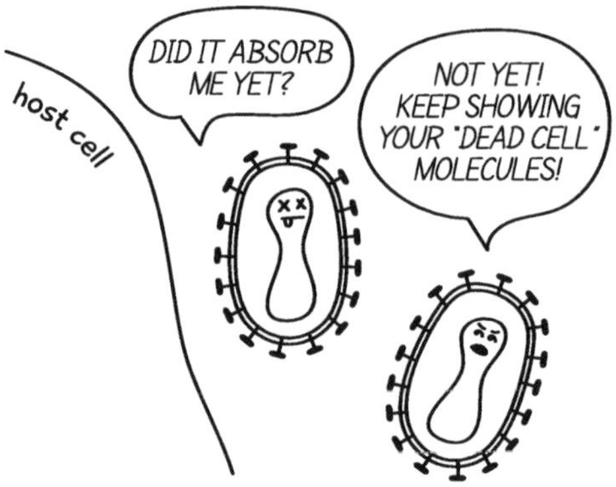

Reminder: Viruses are considered nonliving – this cartoon is just for fun!

Design a medicine to prevent this type of viral mimicry from working. Explain how it does this.

____

____

____

____

REVIEW → **Check & strengthen your understanding by working through this activity.**

Complete these sentences by filling in the missing words.

a. The cell membrane is composed of a phospholipid ____________.

b. We describe the cell membrane using the ____________ mosaic model – which means that it is flexible, allows movement within the membrane, and has different components embedded in it.

c. Small, ____________ molecules can pass freely through the cell membrane. Large and polar molecules require the use of transport ____________.

d. ____________ is a small molecule that regulates the fluidity of the cell membrane at different temperatures.

e. The cell membrane controls what can ____________ and ____________ the cell.

f. Other roles of the cell membrane include cell signaling, cell ____________, and cell adhesion.

© Emma The Teachie Limited 2025-Present. All rights reserved.

QUIZ ➔ **Show what you know by answering each of the questions.**

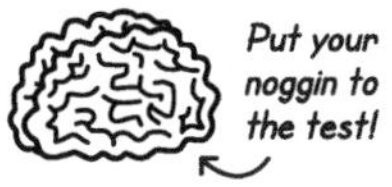

1 Identify each component of the cell membrane shown in the images.

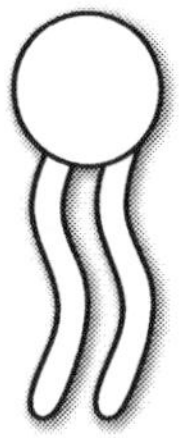

A

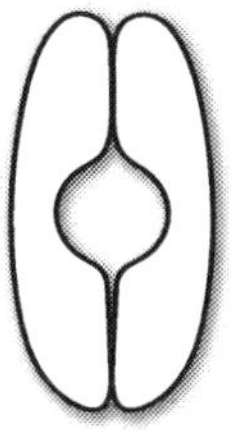

B

C

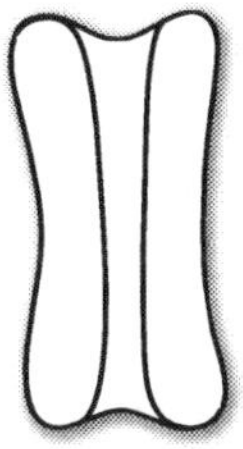

D

2 Which TWO terms are used to describe the cell membrane? (Choose two answers.)

A. Fluid

B. Watery

C. Scattered

D. Mosaic

3 Which of these is NOT a function of the cell membrane?

A. Communicating with other cells

B. Controlling what can enter and leave the cell

C. Adhering to other cells

D. Controlling the cell's activities

4 Which types of molecules can freely pass through the cell membrane?

A. Small polar molecules

B. Large polar molecules

C. Small nonpolar molecules

D. Large nonpolar molecules

5 Which molecule is responsible for maintaining the correct fluidity of the cell membrane at different temperatures?

____

6 How are phospholipids arranged in the cell membrane?

____

____

____

____

7 Which two components are used to transport molecules in and out of the cell membrane?

____

© Emma The Teachie Limited 2025-Present. All rights reserved.

Diffusion and Osmosis

PROCESS ➔ Read the text passage below, then complete these tasks to process the info.

HIGHLIGHT NEW KEYWORDS — SQUIGGLY UNDERLINE OLD KEYWORDS

UNDERLINE ANY DEFINITIONS — ADD ANY RELEVANT DOODLES IN SPACES

You now know that the cell membrane controls the transport of molecules in and out of the cell. But how are these molecules moving through it?

Well, that depends on the molecule and the environment it is in! Let's start with diffusion.

Diffusion

Diffusion is the movement of molecules from an area of high concentration (where there are lots of the molecules) to an area of low concentration (where there are fewer molecules). Molecules move randomly in both directions, however the net movement of molecules is down the concentration gradient.

BEFORE

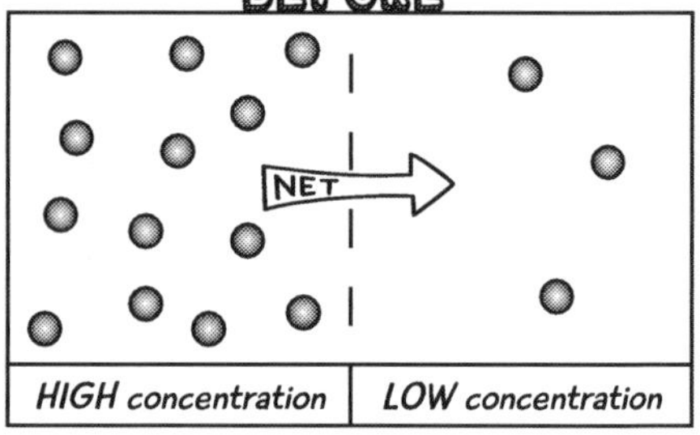

For example, if there were lots of oxygen molecules outside the cell, but not very many inside the cell, then the overall movement would be oxygen molecules moving into the cell.

When the concentration is equal, we say there is equilibrium. We can also call this dynamic equilibrium, as molecules will continue to move at random in both directions, but there is no further net movement.

AFTER

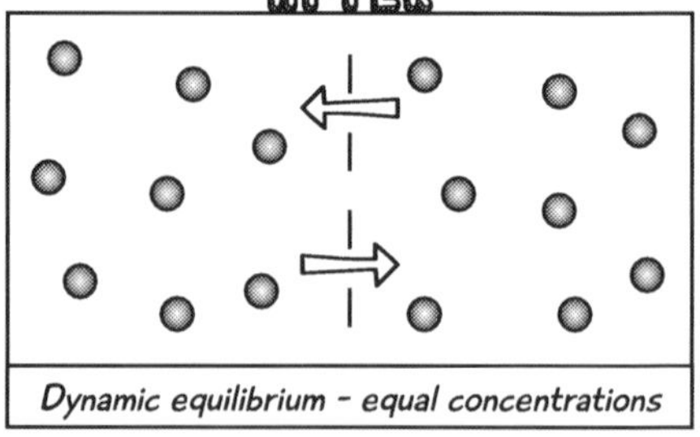

Diffusion is a passive process — this means it doesn't require any extra energy to move the molecules.

© Emma The Teachie Limited 2025-Present. All rights reserved.

Osmosis

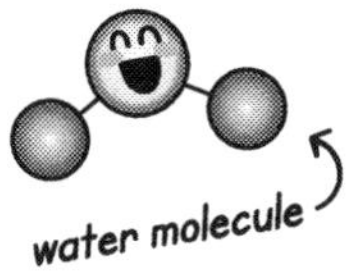

Osmosis is a special type of diffusion — it is the diffusion of water!

Water molecules move by osmosis from an area of high water concentration to an area of low water concentration, across a semipermeable membrane.

Semipermeable means that only specific molecules can pass through — in this case, we're talking about water! Semipermeable membranes may also be called partially permeable, or selectively permeable. The cell membrane is like this. Water molecules can travel directly through it. Sometimes channel proteins are used to speed up the rate of osmosis.

As the water molecules are moving down their concentration gradient, no extra energy is required — so, osmosis is a form of passive transport.

We can also describe the movement of water by osmosis in terms of solute concentration.

Solutes are solids that can dissolve in liquids (these liquids are known as solvents). Salt and glucose are examples of solutes. They are solid — but when we add them to water, they dissolve into it, creating a solution.

If a solution has a high water concentration, it means it has a low solute concentration. And if a solution has a low water concentration, it means it has a high solute concentration.

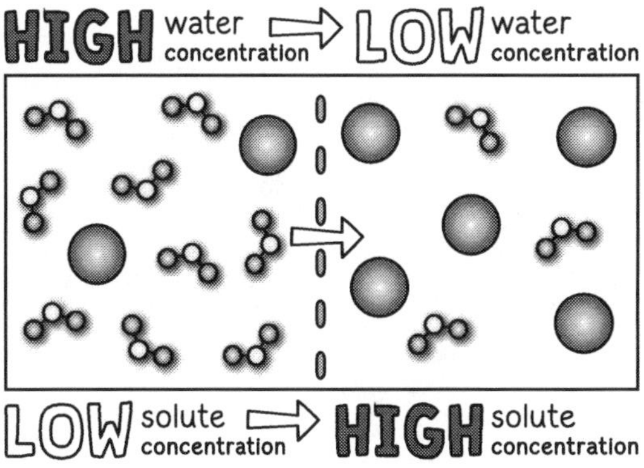

So, we can say that water moves from areas of higher solute concentration to areas of lower solute concentration.

Just like in regular diffusion, osmosis will continue until dynamic equilibrium is reached.

© Emma The Teachie Limited 2025-Present. All rights reserved.

ACTIVATE → Complete the activities to apply what you've learned from the text passage.

1 Get a glass of warm water. Add a single drop of food coloring and observe. What happens to the dye over time? Draw diagrams of what you see happening and add an explanation.

Include labels for your diagrams!

2 Complete these diagrams to show how diffusion works.

- In the "BEFORE" diagram, decide which side is high or low concentration.
- Then, draw an arrow to show the NET direction of movement.
- In the "AFTER" diagram, draw the molecules to show them at equilibrium.

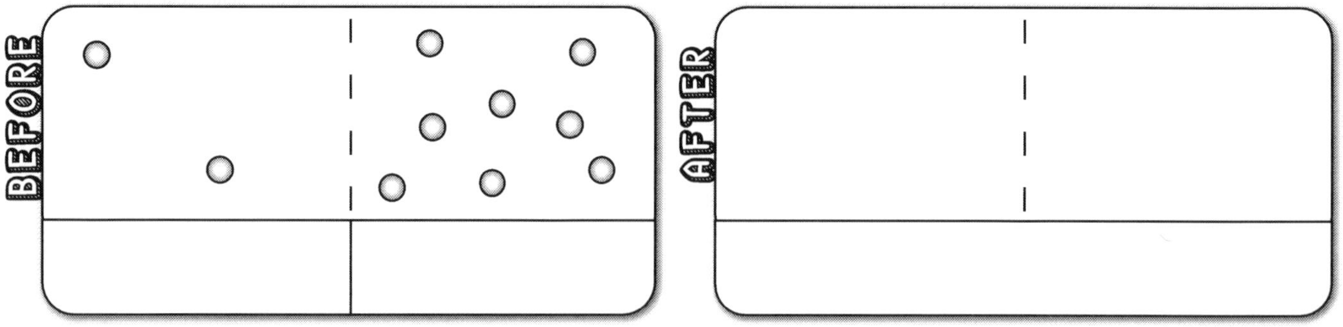

© Emma The Teachie Limited 2025-Present. All rights reserved.

3 For each of these beakers of solution, work out the missing percentage. Then, draw an arrow into or out of the tubing to show the movement of water. If there is no net movement of water, then write this.

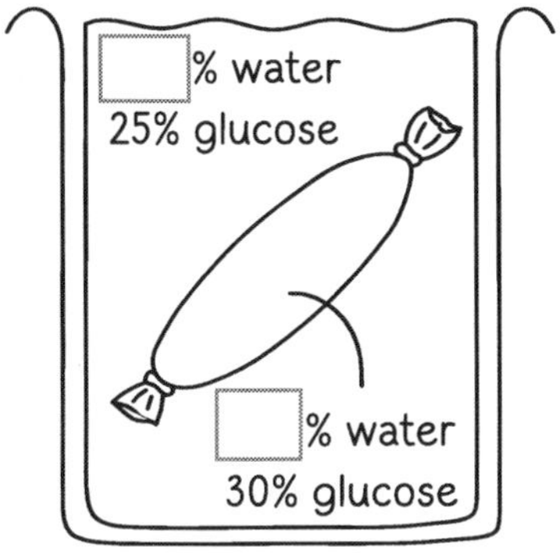

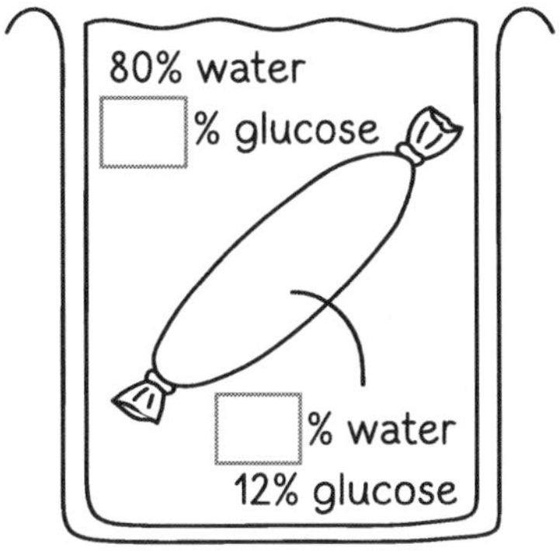

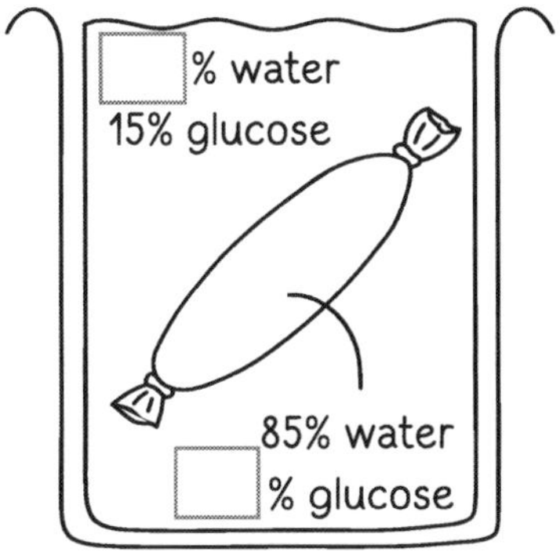

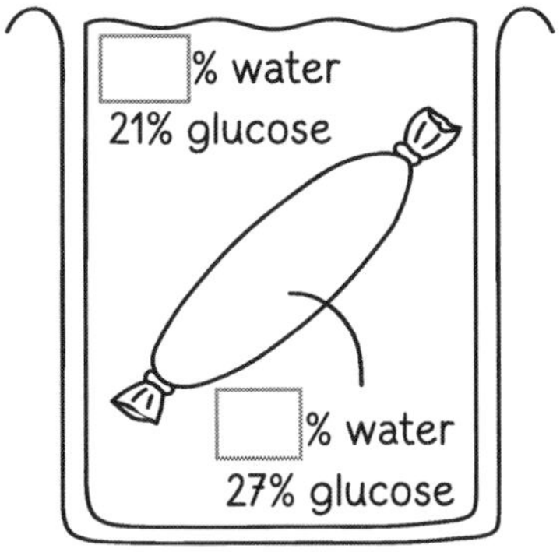

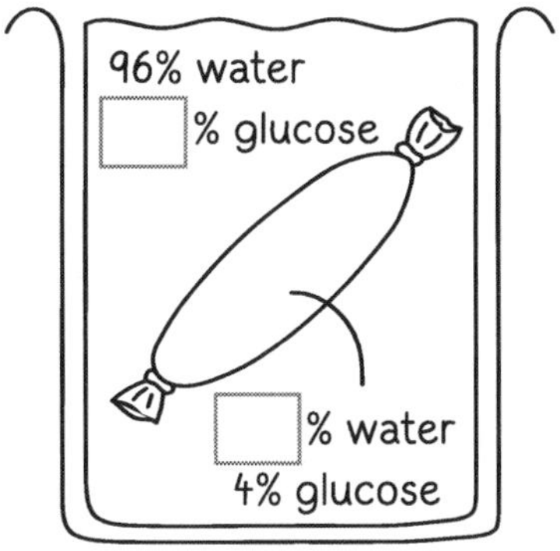

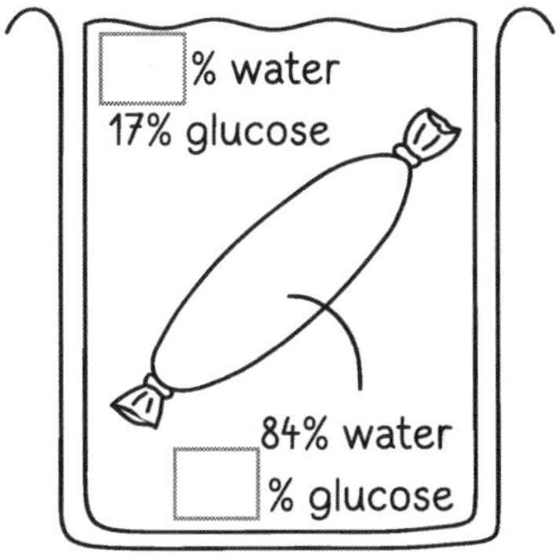

4 Let's look at how the concentration of the external environment can affect animal cells and plant cells. Draw lines to match the solution to its effect on an animal cell, a plant cell, and the correct explanation of why this happens.

SOLUTION TONICITY | **EFFECT ON RED BLOOD CELLS** | **EFFECT ON PLANT CELL** | **WHY THIS HAPPENS**

Isotonic - it is equal in concentration to the inside of the cell

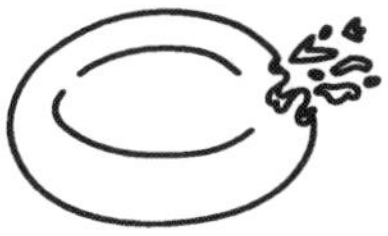

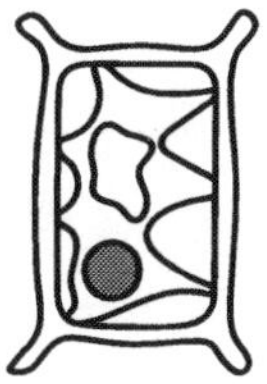

The cell has gained water, as the solute concentration is higher inside the cell. The cytoplasm has swollen. In animal cells, this can cause it to burst. In plant cells, the cell wall prevents this — instead, there is pressure against the cell wall. The cell is turgid — this is its normal state.

Hypotonic - it has a lower solute concentration than the inside of the cell

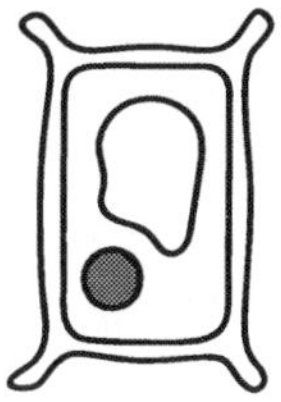

The cell has neither gained nor lost water, as the solute concentrations are equal inside and outside. The animal cell is normal in shape. The plant cell has no pressure exerted against the cell wall and is said to be flaccid.

Hypertonic - it has a higher solute concentration than the inside of the cell

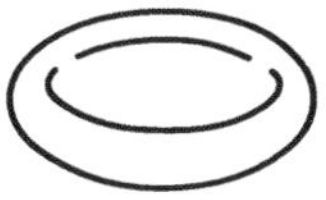

The cell has lost water, as the solute concentration was higher outside the cell. The animal cell will shrivel up and change shape. The plant cell cytoplasm and cell membrane pulls away from the cell wall, and the vacuole shrinks. The cell is said to be plasmolyzed.

© Emma The Teachie Limited 2025-Present. All rights reserved.

EXTEND → **Let's go deeper on this topic and extend your thinking — like a scientist!**

Have you heard of water poisoning? This is a real and potentially fatal situation. It happens when a person drinks very large quantities of water. While uncommon, there have been cases of marathon runners and people doing water-drinking contests or challenges that died from water poisoning.

Explain why drinking very large amounts water in a short period of time could make a person ill.

No need to research – use what you have learned about osmosis.

REVIEW → **Check & strengthen your understanding by working through this activity.**

A student has written notes about diffusion and osmosis. Unfortunately they have made a few mistakes! You are going to help them. Find and underline their mistakes, then rewrite their summary with corrections.

Diffusion is the process of molecules moving from an area of lower concentration to an area of higher concentration. As it is down a concentration gradient, it is an active process.

When the concentration is equal, the molecules stop moving. This is called dynamic momentum.

Osmosis is a special type of diffusion. It is the movement of water molecules, from a high concentration of water to a low concentration of water, across a permeable membrane.

We can also say that water moves from a low solute concentration to a high solute concentration.

© Emma The Teachie Limited 2025-Present. All rights reserved.

QUIZ **Show what you know by answering each of the questions.**

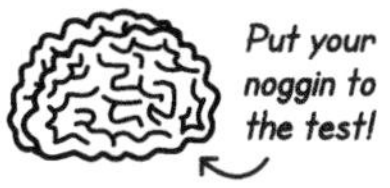

1 **Identify the net direction of movement of the black molecule (draw an arrow).**

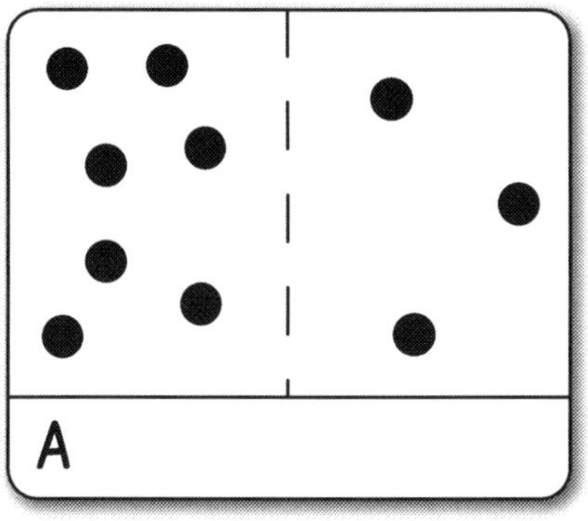

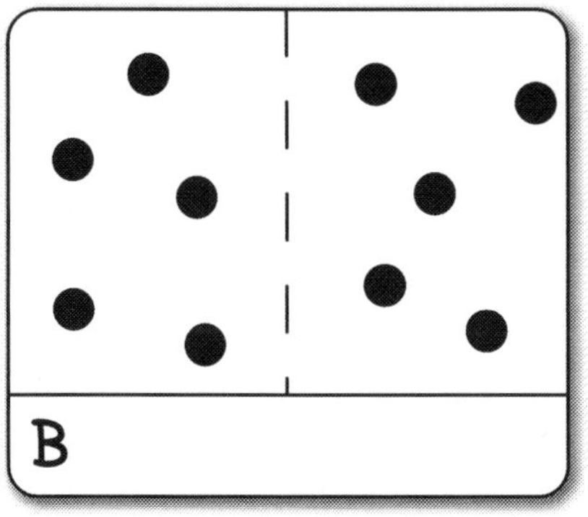

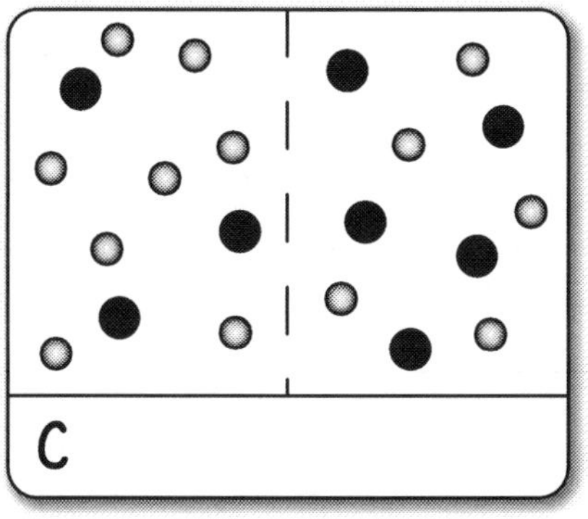

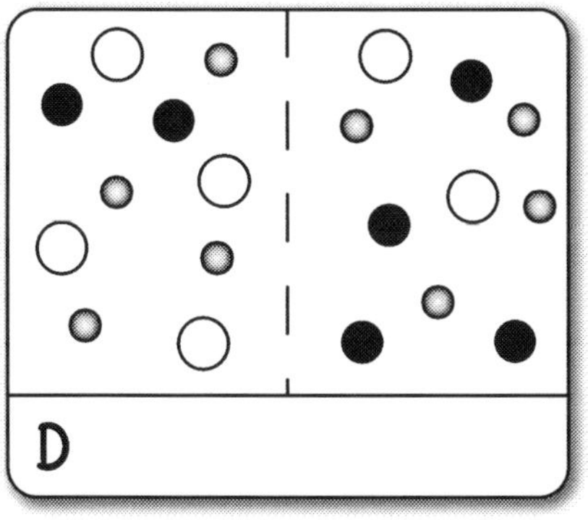

2 **Which of these best describes diffusion?**

A. The net movement of molecules from high to low concentration

B. The net movement of water against a concentration gradient

C. The net movement of molecules from low to high concentration

D. The net movement of molecules across a semipermeable membrane

3 **For osmosis to occur, what is required? There may be more than one answer.**

A. A difference in concentration

B. Extra energy

C. Equilibrium

D. A semipermeable membrane

4 **In osmosis, the net movement of water is...**

A. From low solute to high solute concentration

B. From high solute to low solute concentration

C. From low water to high water concentration

D. None - there is no movement

5 **What is it called when there are equal concentrations of a molecule on both sides of a semipermeable membrane?**

____

6 **What is the name of the semipermeable membrane found in cells?**

____

7 **Explain what happens if a red blood cell is put in a highly concentrated glucose solution.**

____

____

____

© Emma The Teachie Limited 2025-Present. All rights reserved.

Active Transport

PROCESS ➔ **Read the text passage below, then complete these tasks to process the info.**

HIGHLIGHT NEW KEYWORDS — SQUIGGLY UNDERLINE OLD KEYWORDS

UNDERLINE ANY DEFINITIONS — ADD ANY RELEVANT DOODLES IN SPACES

Diffusion and osmosis are both forms of passive transport. Now we're going to learn about their counterpart: active transport!

As you may have guessed, active transport requires the input of extra energy. This is because active transport moves molecules against their concentration gradient, from low concentration to high concentration.

Active transport always occurs across a cell membrane. That's because it requires carrier proteins to move the molecules.

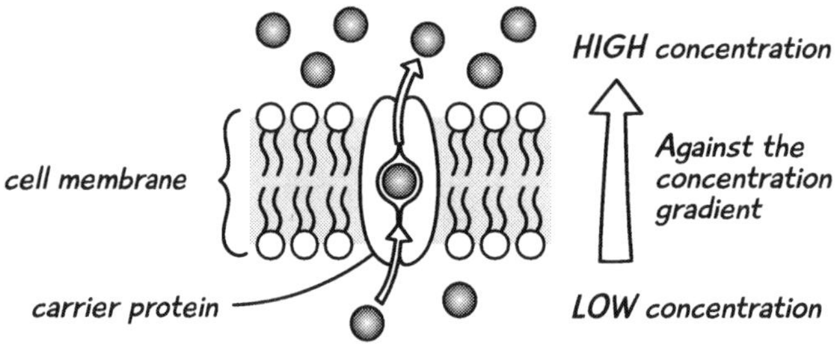

So, why would cells need to move molecules against their concentration gradients if it requires energy?

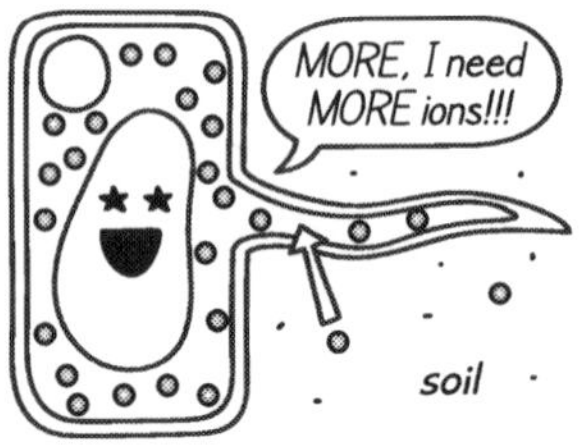

Let's consider the root hair cell. As you know, plants require mineral ions to grow and stay healthy. They absorb these in the roots from the soil. If the soil is low quality and there aren't many mineral ions in it, the cell must be able to absorb them anyway. It does this by using active transport to move the ions against their concentration gradient.

Another example is in the sodium-potassium pump. This is found in all of your cells. We won't go into details, but this pump moves sodium ions and

© Emma The Teachie Limited 2025-Present. All rights reserved.

potassium ions against their concentration gradients to maintain a careful balance.

Now, let's consider what happens if a really big molecule needs to enter or exit the cell. We call these large molecules "macromolecules" (macro = large.)

They may not be able to fit through the cell membrane or its transport proteins. Instead, these molecules are transported in and out of the cell by endocytosis and exocytosis. Both of these processes require extra energy, so they are also a form of active transport.

Endocytosis

In endocytosis, the cell membrane takes in macromolecules by folding in on itself and forming new vesicles made from the cell membrane.

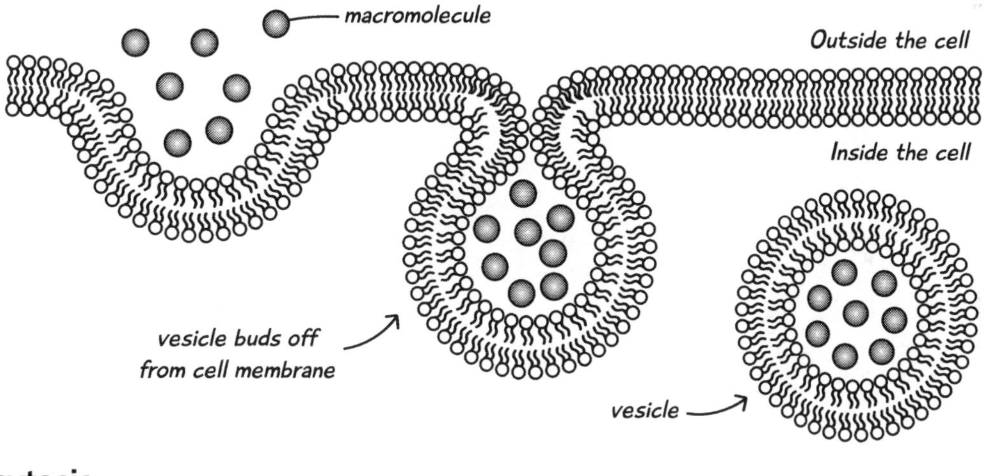

Exocytosis

In exocytosis, the cell membrane releases large molecules. First, an internal vesicle containing macromolecules will move to the cell membrane and fuse with it. Then, the macromolecules are released to the extracellular space.

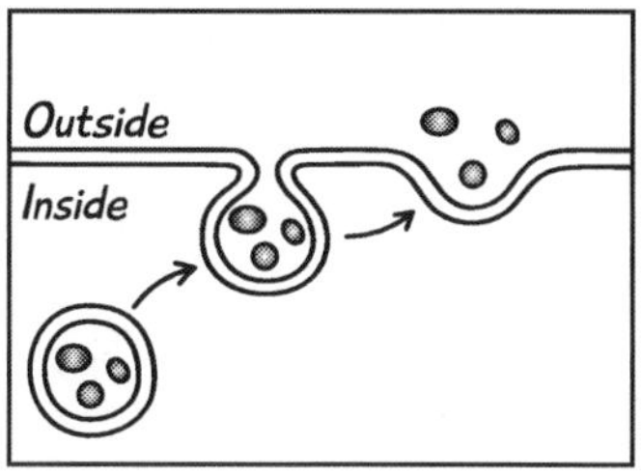

TOP TIP: To remember which is which, think of "en" as 'IN" and "exo" as "EXIT".

© Emma The Teachie Limited 2025-Present. All rights reserved.

ACTIVATE → **Complete the activities to apply what you've learned from the text passage.**

1 Look at each of these pictures, and work out if it is showing active or passive transport. Write your answer in the box below each picture.

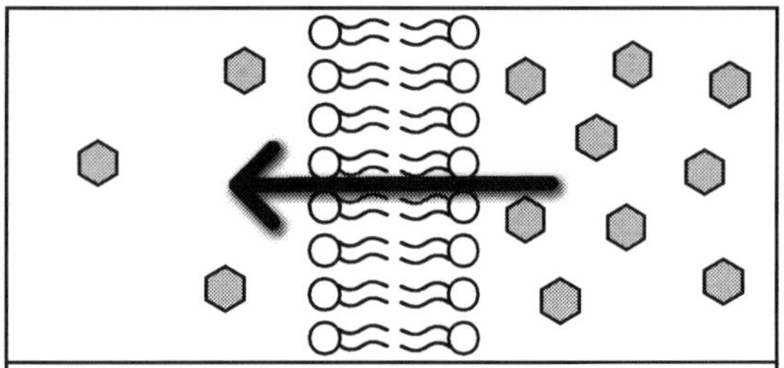

A

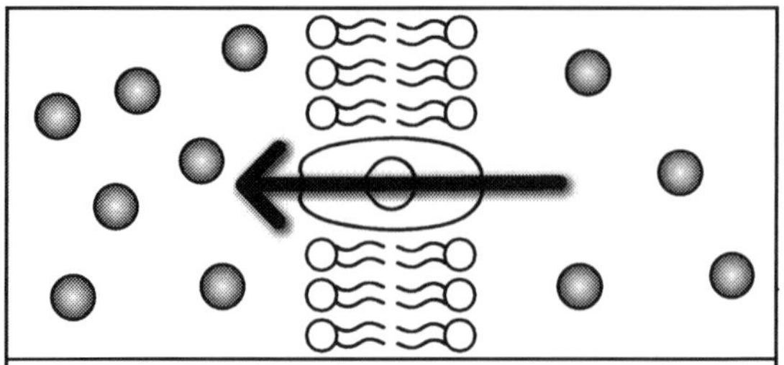

B

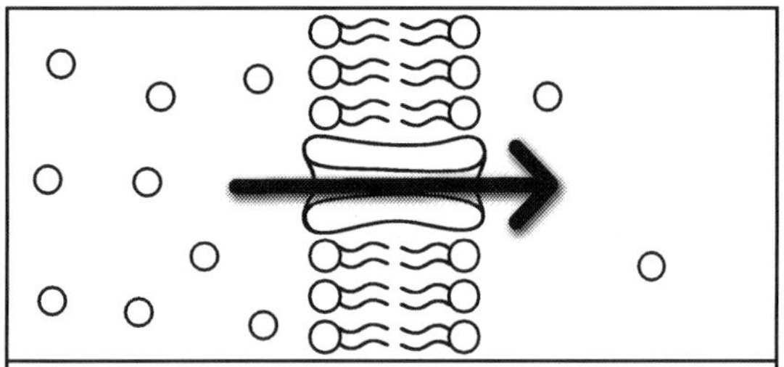

C

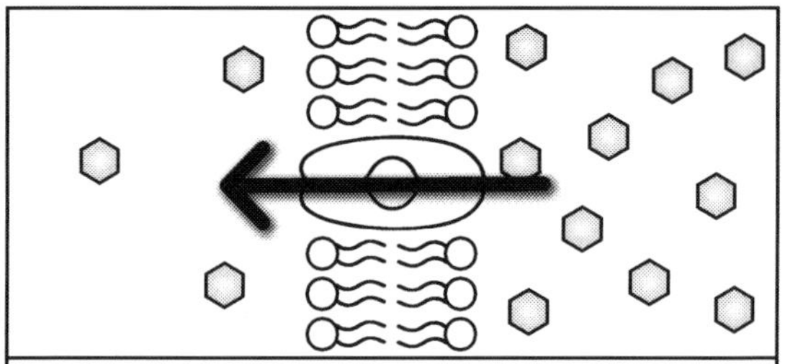

D

2 Both of these diagrams show molecules being moved from a region of low concentration to a region of high concentration. However, neither of them accurately depicts active transport. Explain why for each diagram.

DIAGRAM 1

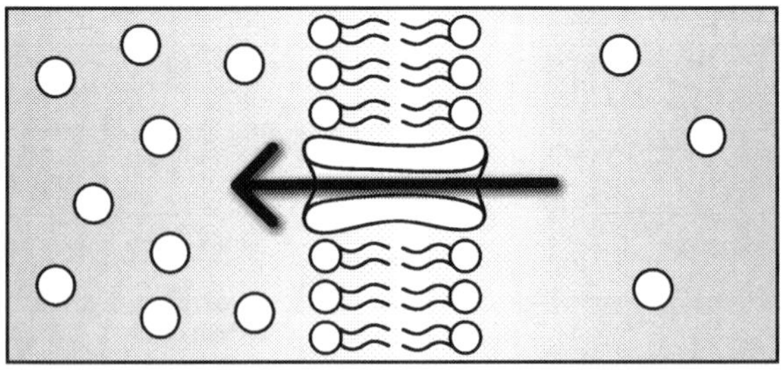

EXPLANATION

DIAGRAM 2

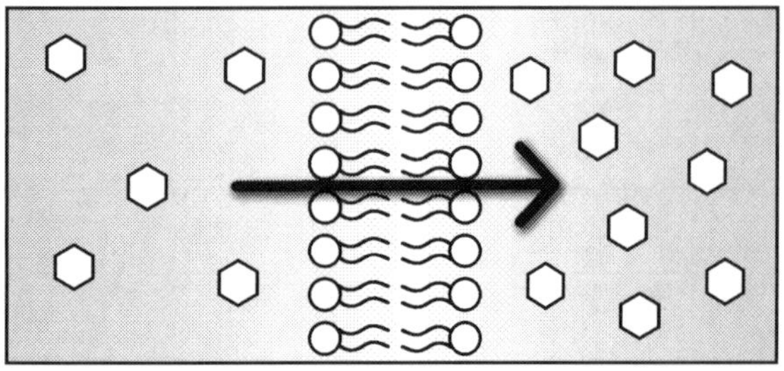

EXPLANATION

© Emma The Teachie Limited 2025-Present. All rights reserved.

3 Create your own simple storyboard to show the process of endocytosis taking place.

4 It's time to compare all of the cell transport processes you've learned about! For each statement, decide if applies to diffusion, osmosis, or active transport. If it applies to two of them, put it in the overlapping part of their circles. If it applies to all three, then put it in the center.

1. From high to low concentration
2. Movement of water
3. Requires energy
4. Movement of molecules
5. From low to high concentration
6. Needs a semipermeable membrane
7. Sometimes uses channel proteins
8. Only uses carrier proteins
9. Occurs naturally in cells
10. A passive process
11. Can move gases
12. Can move solutes

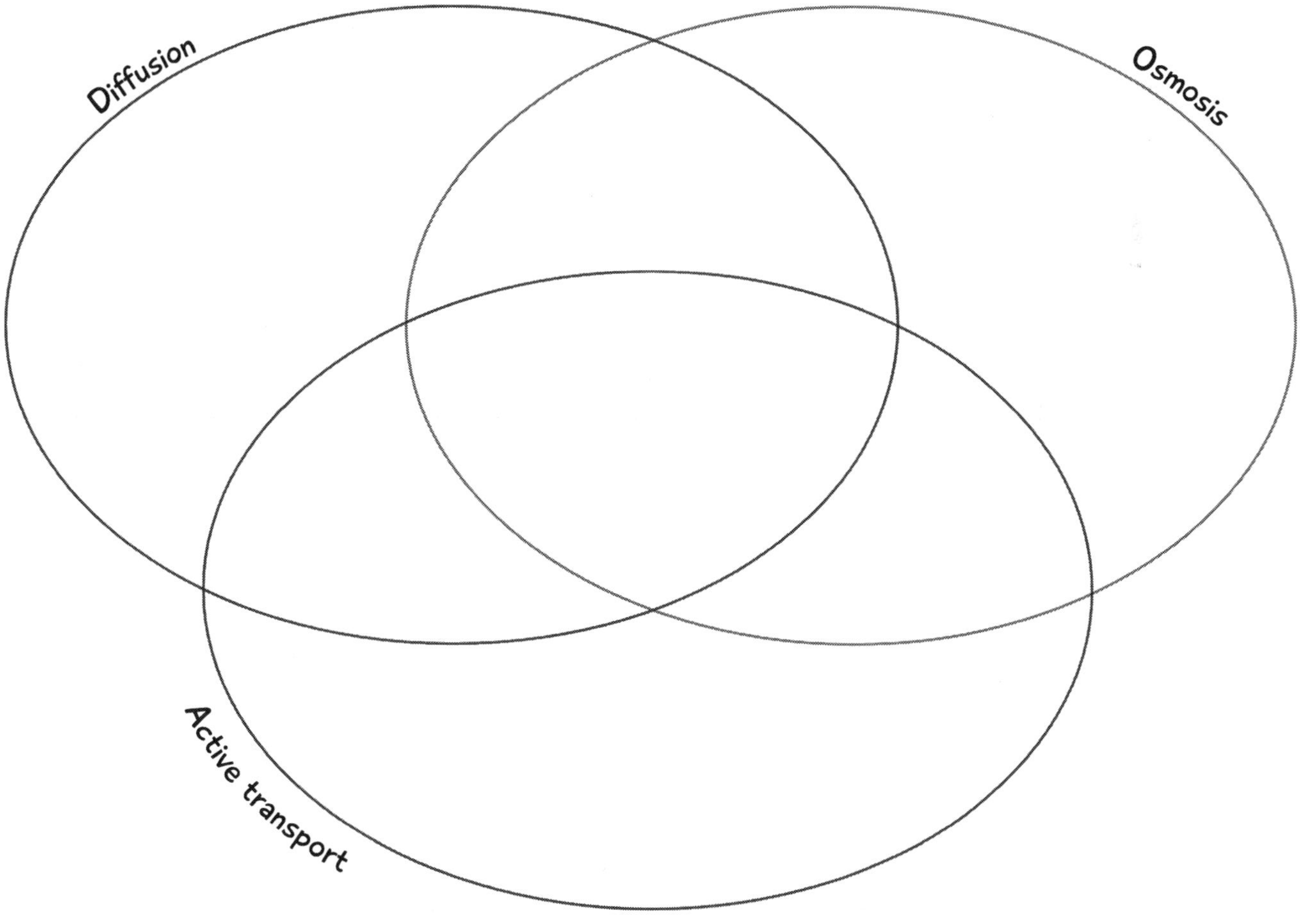

© Emma The Teachie Limited 2025-Present. All rights reserved.

EXTEND ➜ **Let's go deeper on this topic and extend your thinking — like a scientist!**

Create your own analogy for active transport! Here is an example: "Active transport is like carrying groceries uphill into your house instead of letting them roll downhill." Write and illustrate your analogy.

REVIEW ➜ **Check & strengthen your understanding by working through this activity.**

Decide if each statement is TRUE or FALSE — circle T or F. If you think it is false, explain why.

ACTIVE TRANSPORT...	TRUE OR FALSE	If you circled false, why is it false?
1. ...moves molecules from a low to high concentration.	T / F	
2. ...does not require any extra energy.	T / F	
3. ...is a special type of diffusion.	T / F	
4. ...always requires a cell membrane.	T / F	
5. ...uses channel proteins to move molecules.	T / F	
6. ...uses carrier proteins to move molecules.	T / F	

 © Emma The Teachie Limited 2025-Present. All rights reserved.

QUIZ ➔ **Show what you know by answering each of the questions.**

1 Identify each type of cell transport shown in the images.

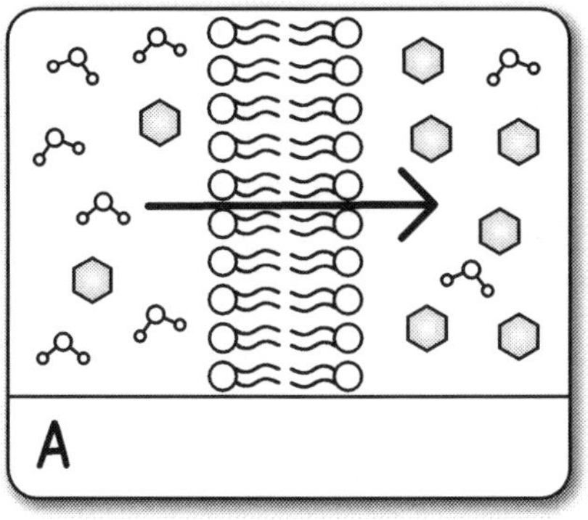

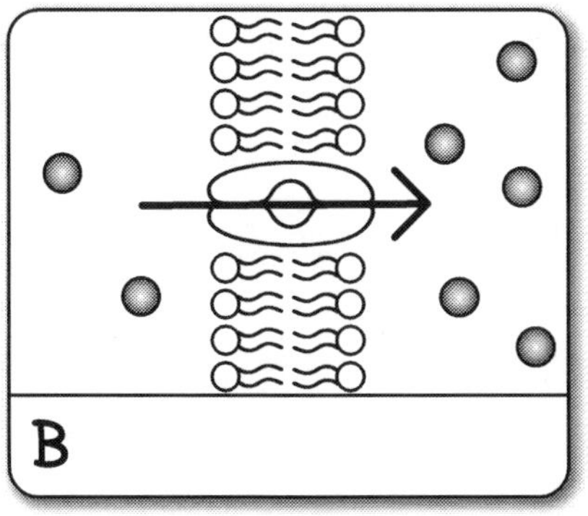

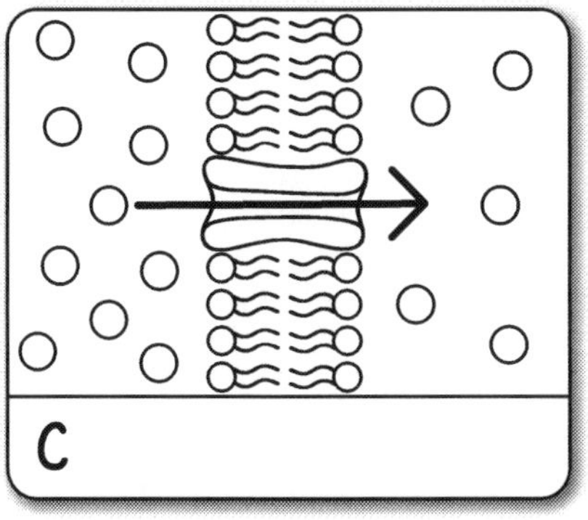

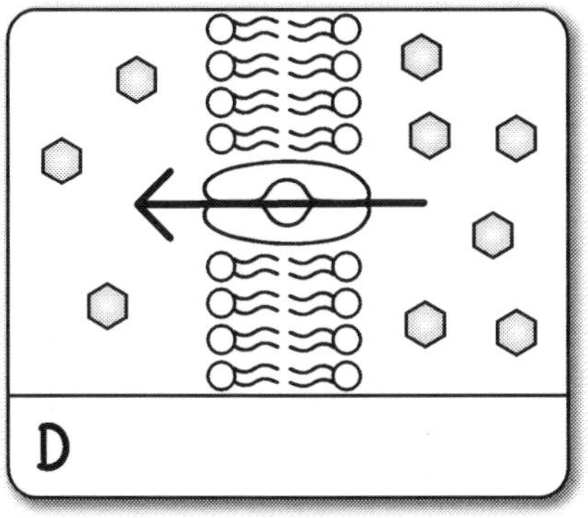

2 Root hair cells use active transport to:

A. Take in mineral ions from the soil

B. Perform photosynthesis

C. Make proteins

D. Absorb oxygen

3 Which of these is NOT true for active transport?

A. It requires extra energy

B. It is the slowest cell transport process

C. It requires carrier proteins

D. It occurs against a concentration gradient

4 Which cell transport process moves water molecules down their concentration gradient?

A. Diffusion

B. Active transport

C. Osmosis

D. Endocytosis

5 Which process brings macromolecules into the cell?

____

6 Which cell structure is always involved in active transport?

____

7 Why does active transport require extra energy?

____

____

____

© Emma The Teachie Limited 2025-Present. All rights reserved.

Photosynthesis

PROCESS → **Read the text passage below, then complete these tasks to process the info.**

HIGHLIGHT NEW KEYWORDS — SQUIGGLY UNDERLINE OLD KEYWORDS

UNDERLINE ANY DEFINITIONS — ADD ANY RELEVANT DOODLES IN SPACES

Nearly all life on Earth depends on photosynthesis. This vital reaction uses light energy from the Sun to convert carbon dioxide and water into glucose and oxygen.

carbon dioxide + water —light→ glucose + oxygen

Plants, algae, and some bacteria use photosynthesis to make their own food (glucose). They are known as autotrophs. Animals, fungi, and most bacteria need to eat other organisms for their energy. They are known as heterotrophs.

Hungry heterotrophs

So how do plants get the carbon dioxide and water needed for photosynthesis?

THINK: Answer this yourself first — how do plants get carbon dioxide and water?

The air in our atmosphere contains carbon dioxide. It enters the plant through tiny holes in the leaves, called stomata. The carbon dioxide then diffuses into the cells in the leaves. Oxygen can also diffuse out of the cells and exit through the stomata. This movement of gases is called gas exchange.

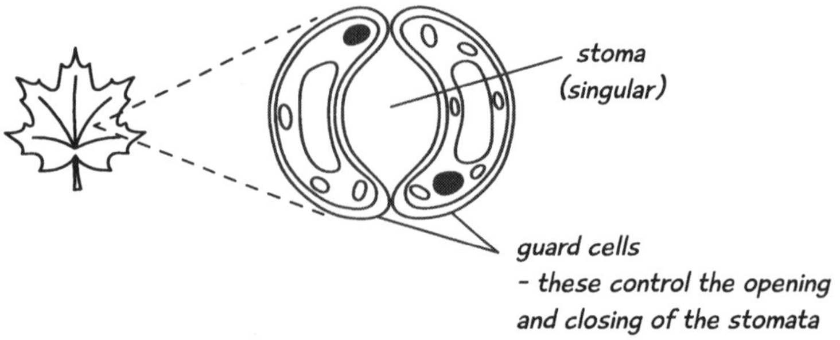

© Emma The Teachie Limited 2025-Present. All rights reserved.

Water enters through the roots of the plants by osmosis. It is transported up through the stem of the plant by the xylem vessel until it reaches the leaves. This is a passive process.

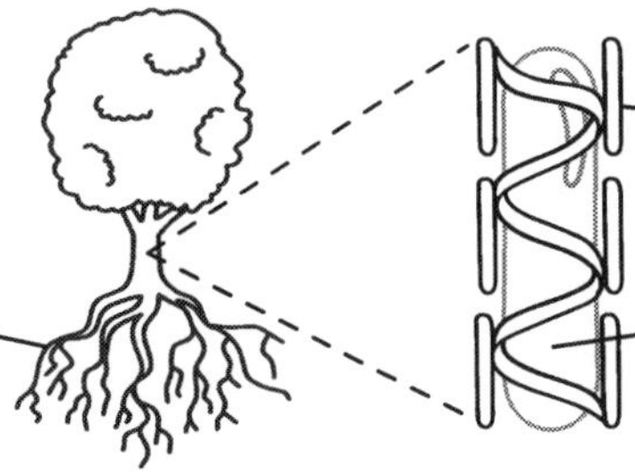

Once the cells in the leaves have carbon dioxide and water, and there is light, then the plant can photosynthesize!

This happens in the chloroplasts. If you recall, these are little green organelles. The reason they are green is because they contain the pigment chlorophyll. There are two types — chlorophyll A and chlorophyll B. These absorb slightly different wavelengths of light. They are best at absorbing red and blue light and worst at absorbing green light. Chlorophyll reflects green light, which is why we see leaves as green!

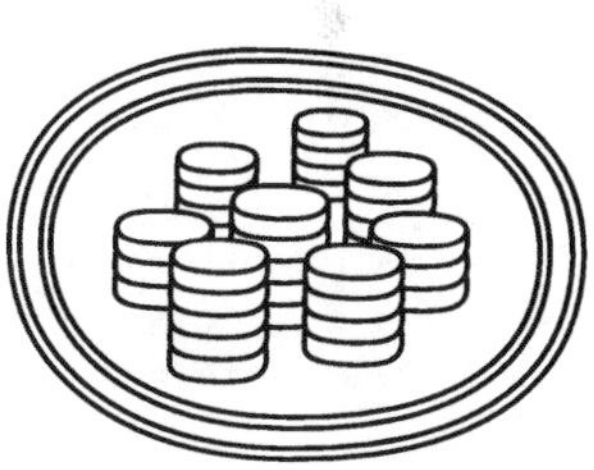

Inside view of a chloroplast

There are other types of pigments found in plants, which is why some leaves are different colors than green, but chlorophyll is the most common.

What does the plant do with the glucose and oxygen created by photosynthesis?

Well, glucose is the food of the plant. It is transported by phloem vessels in both directions, upwards and downwards, to other cells of the plant. This uses some energy. Then, glucose is broken down in cellular respiration to release energy in the form of ATP. This powers the cell's activities!

Some of the oxygen produced in photosynthesis may be used in cellular respiration. But, if there is an excess of oxygen, it will diffuse out of the cells and leave through the stomata.

© Emma The Teachie Limited 2025-Present. All rights reserved.

ACTIVATE → **Complete the activities to apply what you've learned from the text passage.**

1 The plant in the diagram below is photosynthesizing! Fill in the boxes to show the reactants (what goes in) and the products (what comes out) of this reaction.

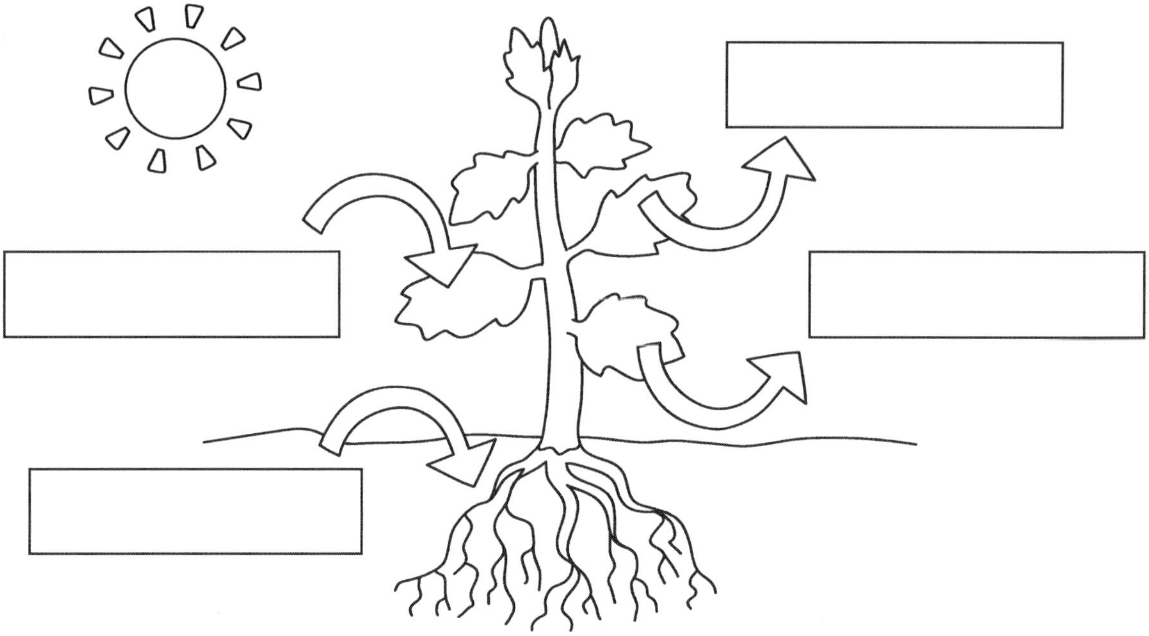

TAKE IT FURTHER: Add your own extra labels to give more detail to this diagram!

2 The diagram below shows the molecules involved in photosynthesis. Look carefully!

- Work out what each molecule is - use the key below.
- Then, write the word equation. Name each molecule with words.
- Then, write the symbol equation. Use chemical symbols for each molecule (e.g., O_2 = oxygen), and then add BIG numbers in front to show how many molecules there are (e.g., $6O_2$).

KEY ● = oxygen ○ = carbon ○ = hydrogen

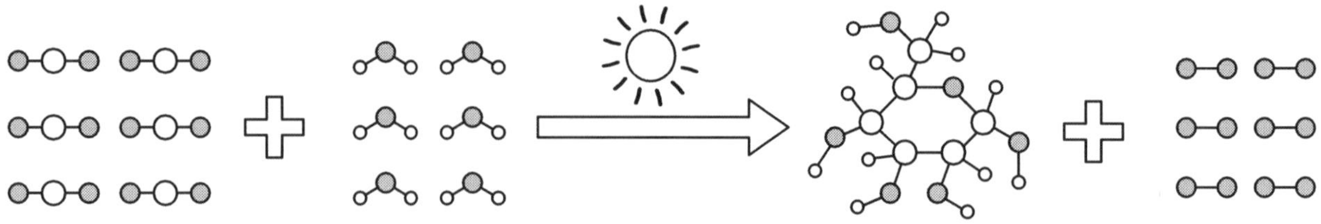

WORD EQUATION

SYMBOL EQUATION

© Emma The Teachie Limited 2025-Present. All rights reserved.

3 Photosynthesis couldn't happen without transport of molecules through the plant. Compare the two transport vessels by answering these three questions for each of them:

XYLEM

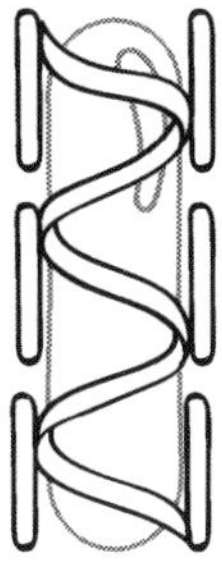

What does it transport?

In which directions can it transport molecules?

Does it require energy to transport molecules?

PHLOEM

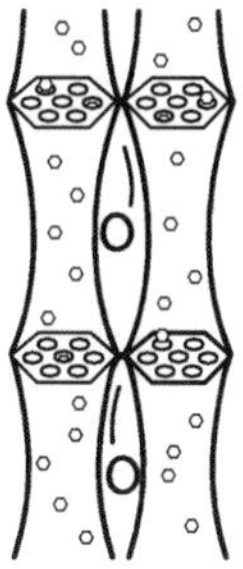

4 Complete the missing words for each step.

A	Water is absorbed from the soil by ________ in the _____ of the plant. Root hair cells have a large surface ____ to increase the rate of water uptake.
B	Water is _____________ up through the stem of the plant to the leaves. This happens in the _____ vessel.
C	Carbon ________ is taken in through small holes in the leaves called ________.
D	The leaves contain green organelles called ______________. They contain the green pigment _____________ which can absorb light. This is used to convert carbon dioxide and water into _______ and ________.
E	Glucose is transported to other parts of the plant by the _______ vessel. It is broken down in the process of cellular respiration to release _______ for the cell's activities.
F	Excess oxygen can ________ out of the stomata in the leaves.

5 Compare autotrophs and heterotrophs by putting a check mark in each row.

CHARACTERISTIC	AUTOTROPH	HETEROTROPH
☆ Makes its own food for energy	☐	☐
☆ Needs to eat other organisms	☐	☐
☆ Includes animals and fungi	☐	☐
☆ Includes plants	☐	☐
☆ Uses photosynthesis or chemosynthesis	☐	☐

© Emma The Teachie Limited 2025-Present. All rights reserved.

EXTEND ➔ Let's go deeper on this topic and extend your thinking — like a scientist!

Let's take a closer look at the stomata — the tiny holes found mostly on the underside of leaves. They can open and close. This is controlled by guard cells. When water enters the guard cells, it increases the pressure and causes it to swell, forming a bean shape. When guard cells lose water, they become flaccid, and the stoma closes.

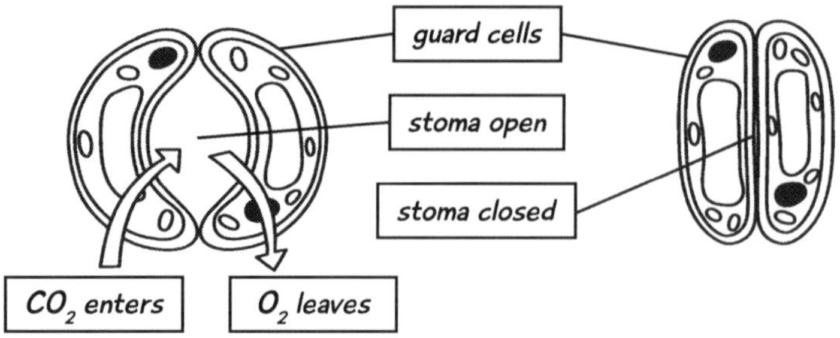

1. What is the purpose of the stomata in photosynthesis?

2. Water vapor can escape from open stomata. Suggest why stomata are able to close.

3. For most types of plants, their stomata are usually open during the day and closed at night. Explain why.

REVIEW ➔ Check & strengthen your understanding by working through this activity.

Find someone to explain the process of photosynthesis to. As you explain it, get them to check off each keyword that you use. If you miss any, try and add another sentence to explain it.

(If no one is around right now, you can record yourself explaining it and share it later!)

PHOTOSYNTHESIS KEYWORD CHECKLIST

☐ Water	☐ Xylem	☐ Chlorophyll	☐ Oxygen
☐ Root hair cells	☐ Phloem	☐ Stomata	☐ Diffusion
☐ Glucose	☐ Chloroplasts	☐ Carbon dioxide	☐ Gas exchange

 © Emma The Teachie Limited 2025-Present. All rights reserved.

QUIZ → **Show what you know by answering each of the questions.**

1 Identify the components/ molecules shown in each image.

A

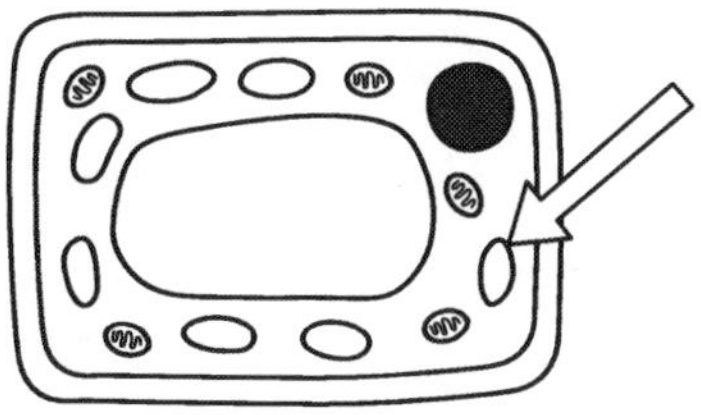

B

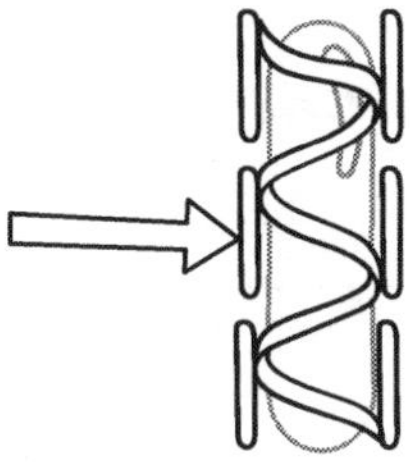

C

$C_6H_{12}O_6$

D

2 Which type of organism can carry out photosynthesis?

A. Autotrophs

B. Heterotrophs

C. Carnivores

D. Herbivores

3 What is the major photosynthetic organ of most plants?

A. Leaves

B. Stems

C. Roots

D. Flowers

4 What is the function of photosynthesis?

A. To absorb light for the plant

B. To produce food for the plant

C. To absorb water for the plant

D. To transport molecules around the plant

5 What is the word equation for photosynthesis?

____

6 What is the name of the green pigment that absorbs light for photosynthesis?

____

7 Describe how a plant obtains the gas needed for photosynthesis. Make sure you name the gas.

____

____

____

© Emma The Teachie Limited 2025-Present. All rights reserved.

Cellular Respiration

PROCESS → **Read the text passage below, then complete these tasks to process the info.**

- HIGHLIGHT NEW KEYWORDS
- SQUIGGLY UNDERLINE OLD KEYWORDS
- UNDERLINE ANY DEFINITIONS
- ADD ANY RELEVANT DOODLES IN SPACES

We need energy to run, think, and even sleep. This energy comes from a process called cellular respiration. This is how cells break down glucose to release energy in the form of ATP.

ATP is a little molecule that acts a bit like a battery! It powers everything a cell does, from dividing, to sending signals, to building new molecules.

There are a few types of cellular respiration.

Aerobic cellular respiration - This is happening in your cells right now (unless you are sprinting while you read this!). It uses oxygen to fully break down the glucose, releasing lots of ATP.

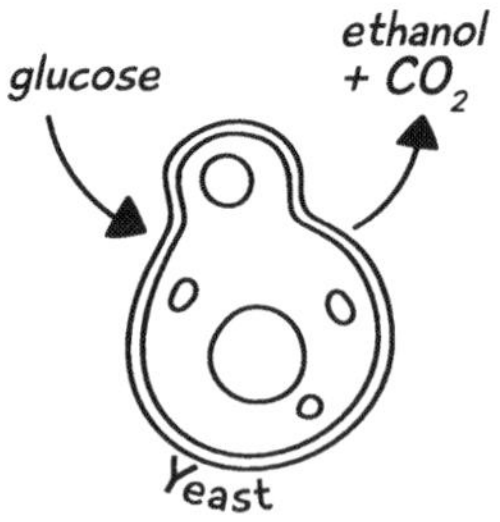

Anaerobic cellular respiration - This happens when there is not enough oxygen available. In animal cells, anaerobic cellular respiration produces lactic acid. In yeast, anaerobic cellular respiration produces ethanol and carbon dioxide. It is also known as fermentation.

Let's look in more detail at aerobic cellular respiration.

Where does it happen?

In animal and plant cells, aerobic cellular respiration takes place in the mitochondria. You might remember mitochondria as the "powerhouses" of the cell — and now you know why! Without them, there would be no energy available for the cell.

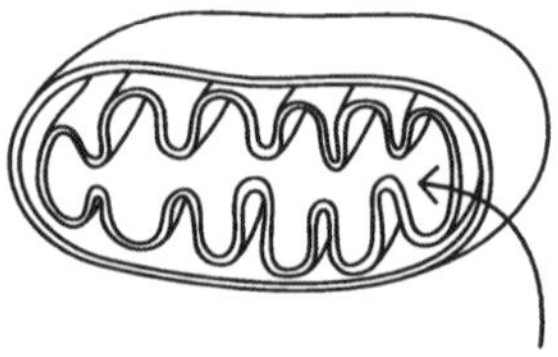

cross section of a mitochondrion

© Emma The Teachie Limited 2025-Present. All rights reserved.

Let's look at what's needed for aerobic cellular respiration:

Glucose - In animals, this comes from our food. In plants, this comes from photosynthesis.

Oxygen - This comes from the air we breathe in. It gets carried by red blood cells from the lungs to respiring cells. In plants, this comes from photosynthesis or from the air that enters through the stomata in the leaves.

Oxygen fully breaks down the glucose, releasing the energy stored in its bonds. This is used to produce lots of ATP. Carbon dioxide and water are also produced as byproducts.

glucose + oxygen → carbon dioxide + water + ATP (energy)

Sometimes you'll see this equation without the ATP/energy part — both are ok!

So what happens to the **carbon dioxide** and **water** that are made during aerobic respiration?

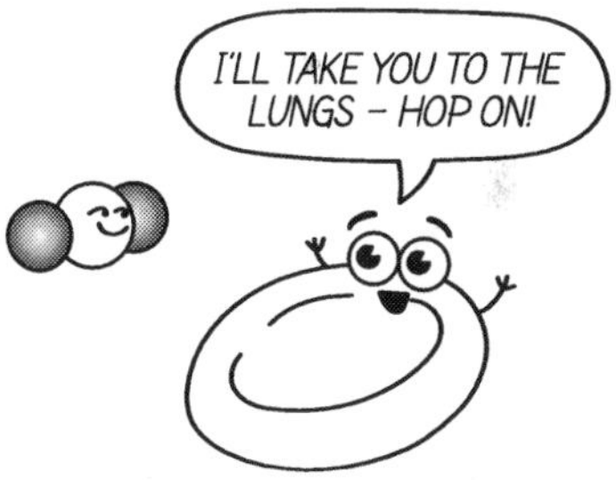

In animals, carbon dioxide diffuses out of the respiring cells and into the blood stream. It is transported to the lungs and breathed out. The water produced can be used if needed, or it can be released from the body in various ways — sweating, urinating, and in our breath. If you've ever been outside on a very cold day, you might have seen the water vapor in your breath condense as it makes contact with the cold air!

In plants, carbon dioxide can be reused in photosynthesis or released through the stomata. The water can also be used for photosynthesis, or it can leave through stomata in the leaves as water vapor.

© Emma The Teachie Limited 2025-Present. All rights reserved.

ACTIVATE → **Complete the activities to apply what you've learned from the text passage.**

1 Write the word equations for each type of cellular respiration. For simplicity, do not write ATP/energy in the equation.

	REACTANTS		PRODUCTS
Aerobic respiration	______________	→	______________
Anaerobic respiration in animals	______________	→	______________
Anaerobic respiration in yeast	______________	→	______________

THINK: Compare the reactants and products of aerobic respiration and photosynthesis. What do you notice?

2 Cellular respiration happens in every living cell — from a cell in your big toe to a neuron in your brain! Animal cells, plant cells, bacteria, yeast, fungi — it happens in all of them!

Compare the process of aerobic cellular respiration in animals and plants by completing this table.

	AEROBIC RESPIRATION IN ANIMAL CELLS	*AEROBIC RESPIRATION IN PLANT CELLS*
Where it happens		
How the glucose is obtained		
How oxygen is obtained		
Products		
What happens to the water produced		
What happens to the carbon dioxide produced		

© Emma The Teachie Limited 2025-Present. All rights reserved.

3

Sort the statements into the Venn diagram to show the similarities and differences of aerobic and anaerobic respiration. Do this by writing the statement number in the diagram.

1. Releases energy
2. No oxygen needed
3. Happens in mitochondria
4. Breaks down glucose
5. Requires oxygen
6. Glucose fully broken down
7. Happens in cells
8. Glucose partially broken down
9. Happens in cytoplasm

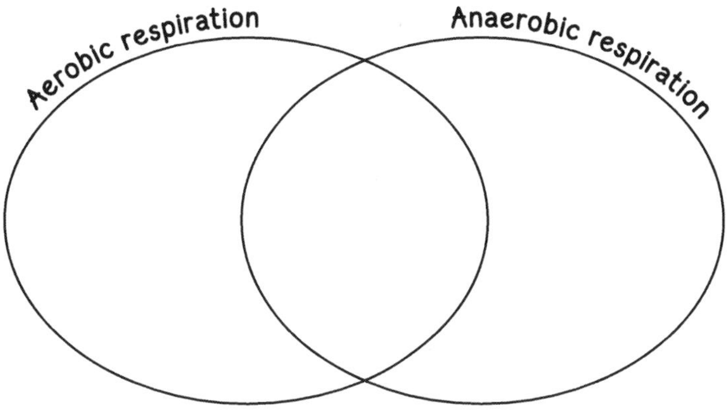

4

Let's compare the process of aerobic cellular respiration with photosynthesis. Decide which process each statement applies to and check the circle to show this. BUT WAIT! There are some bogus statements that don't apply to either process! If you spot one of these, check the "neither" option.

STATEMENT	PHOTOSYNTHESIS	AEROBIC RESPIRATION	NEITHER!
Occurs in chloroplasts	○	○	○
Occurs in the nucleus	○	○	○
Requires oxygen	○	○	○
Generates ATP/energy	○	○	○
Happens all the time	○	○	○
Occurs in mitochondria	○	○	○
Happens during daylight	○	○	○
Requires light energy	○	○	○
Produces lactic acid	○	○	○
Produces glucose and oxygen	○	○	○
Produces water and carbon dioxide	○	○	○
Done by almost all cells	○	○	○
Only done by autotrophs	○	○	○

© Emma The Teachie Limited 2025-Present. All rights reserved.

EXTEND → **Let's go deeper on this topic and extend your thinking — like a scientist!**

Imagine a runner is sprinting at their top speed. As you know, respiration will be happening to produce all of the energy needed for their movement. Let's dig deeper into what is happening.

1. What type of respiration are their muscle cells using? Why?

2. Which product builds up in their muscles, and what effect does this have?

3. Suggest why we cannot sprint for long periods of time.

REVIEW → **Check & strengthen your understanding by working through this activity.**

These are the answers... what is the question?! For each keyword, write a question relating to respiration that would have the keyword as the answer. There may be a few different questions that could have the same answer.

The answer!	The question!
Mitochondria	
Ethanol	
ATP	
Glucose	
Red blood cells	
Carbon dioxide	
Aerobic	

© Emma The Teachie Limited 2025-Present. All rights reserved.

QUIZ Show what you know by answering each of the questions.

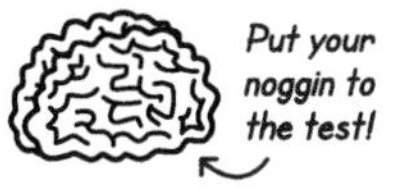

1 Identify each of these molecules as a reactant or a product of respiration.

WATER

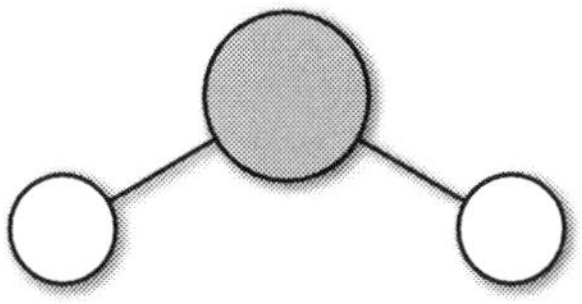

A

GLUCOSE

B

CARBON DIOXIDE

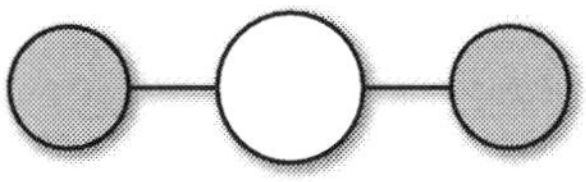

C

OXYGEN

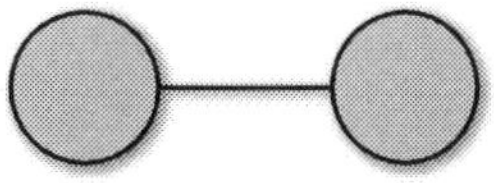

D

2 What is the purpose of cellular respiration?

A. To produce water for the cell

B. To release energy for the cell

C. To use up the glucose taken into the cell

D. To produce oxygen for the cell

3 Where does aerobic cellular respiration happen?

A. Cytoplast

B. Chloroplasts

C. Ribosomes

D. Mitochondria

4 Which product is formed in animals during anaerobic respiration?

A. Ethanol

B. Oxygen

C. Lactic acid

D. Carbon dioxide

5 In plants, which structure allows gases to enter and leave?

____

6 State one way animals release the water produced in aerobic respiration.

____

7 Describe two differences between photosynthesis and aerobic respiration.

____

____

____

____

© Emma The Teachie Limited 2025-Present. All rights reserved.

UNIT 3 REVIEW CROSSWORD

DOWN

1 Molecule in cell membranes that keeps them flexible at different temperatures
2 Gas used by plants for photosynthesis
3 The process plants use to make food using sunlight
5 State reached when the concentration of a substance is balanced
6 Main component of the cell membrane, with a water-loving head and water-hating tail
8 Protein that changes shape to move molecules through the membrane
10 Green pigment in chloroplasts that captures light energy
11 Movement of particles from high to low concentration
13 Process where a cell releases materials through vesicles
15 Membrane protein that acts like a tunnel to help substances cross the cell membrane
16 A function of the cell membrane that helps cells stick together
17 Plant tissue that carries water from roots to leaves
20 Energy-carrying molecule of the cell

ACROSS

4 Type of respiration that doesn't require oxygen
7 A thin barrier that controls what enters and exits the cell
9 Movement of molecules against their concentration gradient, requiring energy
12 Cellular process that breaks down glucose to release energy
14 Diffusion of water through a selectively permeable membrane
18 An organism that makes its own food
19 Product of photosynthesis that is transported around the plant and used for energy
21 Tiny openings on leaves where gases enter and exit

© Emma The Teachie Limited 2025-Present. All rights reserved.

UNIT 4
CELL GROWTH AND REPRODUCTION

© Emma The Teachie Limited 2025-Present. All rights reserved.

The Cell Cycle

PROCESS ➔ **Read the text passage below, then complete these tasks to process the info.**

HIGHLIGHT NEW KEYWORDS — SQUIGGLY UNDERLINE OLD KEYWORDS

UNDERLINE ANY DEFINITIONS — ADD ANY RELEVANT DOODLES IN SPACES

Almost all of the cells in our body don't live as long as we do. They get constantly replaced, making new cells to replace old, worn out cells.

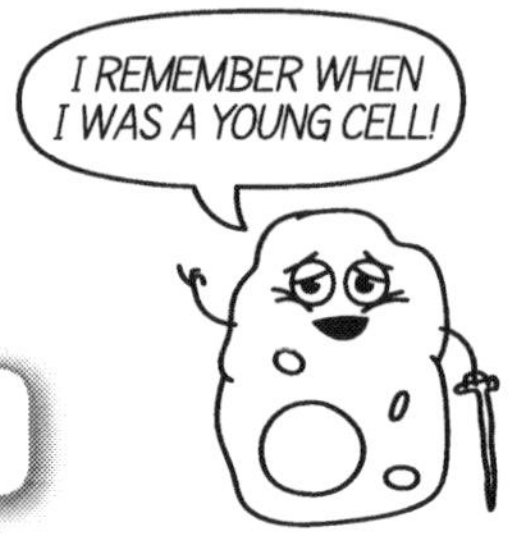

Did you know? Some cells in the eyes and brain are with us from birth to death!

To make new cells, the original cell has to grow and produce new organelles before it is ready to divide into two cells. The growth phase is known as interphase. And the division phase is known as mitosis. Together, they form the cell cycle. It is a cycle, because cells continually move through these stages.

Interphase

This the longest phase of the cell cycle. It can be broken down into three growth stages — G1, S, and G2.

Some cells enter a phase called **G0** before G1 — This is a resting phase for when the conditions for cell division are not favorable, e.g., limited resources.

G1: This is the first growth phase. The cell starts to grow, synthesizing new proteins and organelles, e.g. mitochondria, ribosomes, lysosomes, etc. This is important in preparation for cell division, as two daughter cells are produced during mitosis, and each one will require its own organelles.

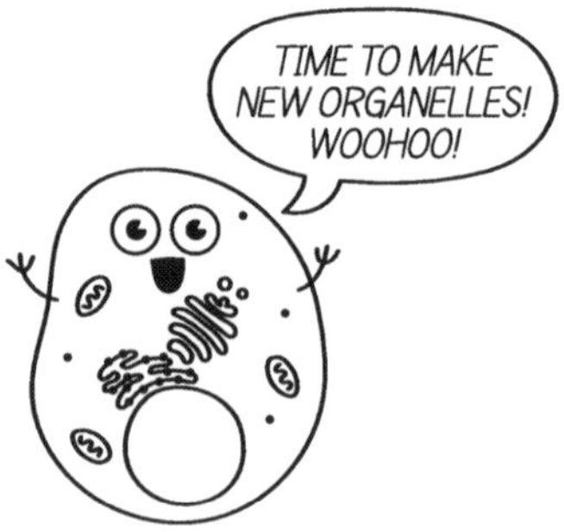

S: This is known as the synthesis phase. During the S phase, the cell replicates its DNA. The DNA is unwound and replicated using special enzymes and

© Emma The Teachie Limited 2025-Present. All rights reserved.

proteins. This is important as each new cell will require its own genetic material to control the cell's activities.

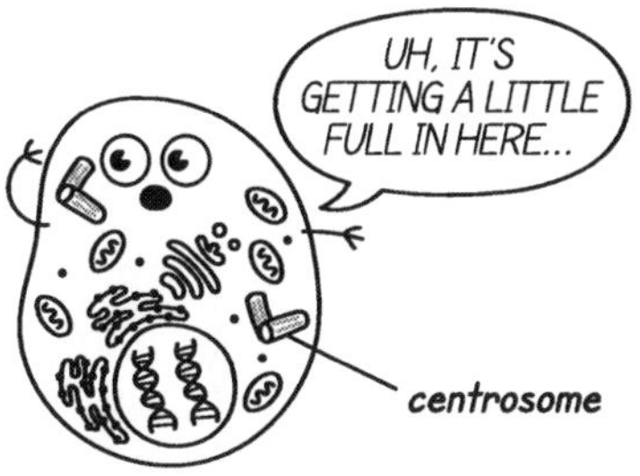

G2: This is the second growth phase. The cell continues to grow, synthesizing the proteins that will be needed for cell division, e.g., centrosomes and spindle fibers. Now the cell is ready to divide!

Mitosis

Body cells divide by mitosis. You're going to learn more details of this in the next section, but here's the overview: During mitosis, the two copies of DNA are carefully separated, moving to opposite ends of the cell.

Cytokinesis

The cell then splits into two by dividing its cytoplasm — this is known as cytokinesis. The result is two genetically identical daughter cells.

genetically identical daughter cells

What next?

The cells return to interphase after cytokinesis, once again carrying out their specific functions and preparing for the next round of cell division.

How often they divide depends on their cell type and if any damage is experienced. For example, epidermal skin cells undergo mitosis frequently, as this part of the skin is constantly shed and replaced.

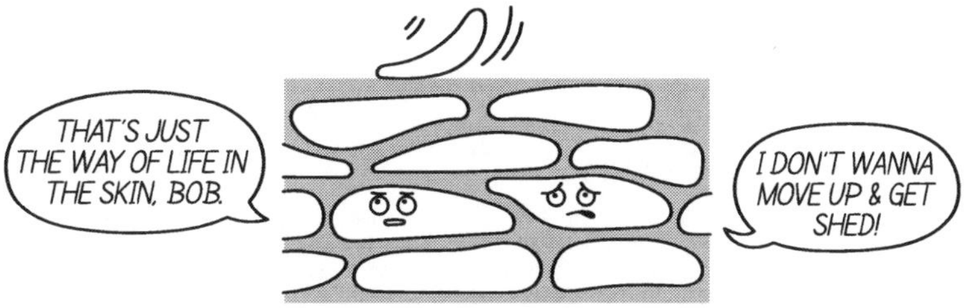

© Emma The Teachie Limited 2025-Present. All rights reserved.

ACTIVATE → Complete the activities to apply what you've learned from the text passage.

1 The pictures below show the cell cycle — but they aren't finished! Only G0 and Mitosis have been done. The rest of the cell cycle pictures are missing details...

- Add doodles to show what is happening in each stage.
- Write what is happening beside each cell.
- In the labels, add the names of the cell cycle (G0 has been done for you.)

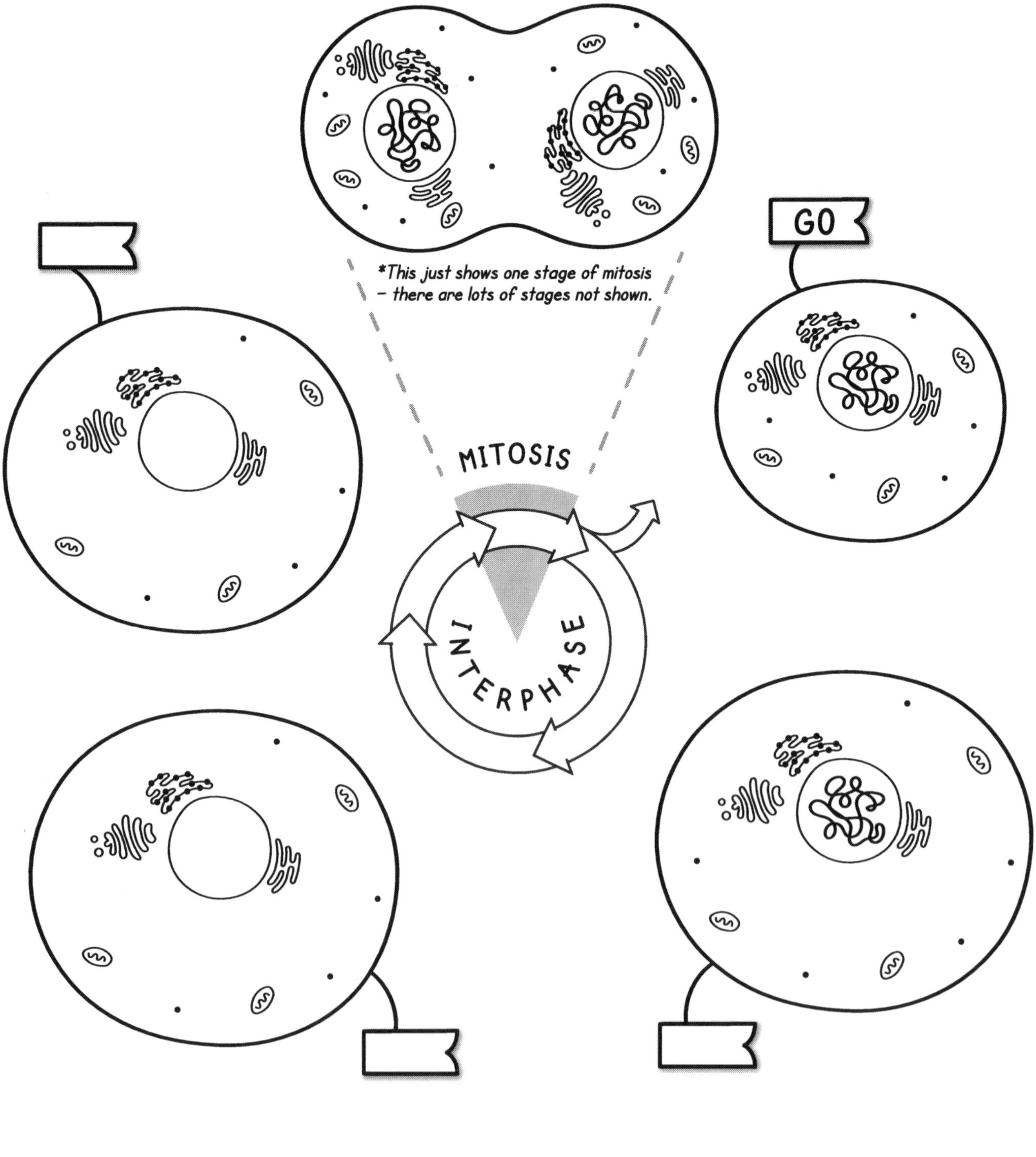

© Emma The Teachie Limited 2025-Present. All rights reserved.

2 Decide which part of the cell cycle each process belongs to — interphase or mitosis? Check your answer.

PROCESS	INTERPHASE	MITOSIS
DNA is copied	☐	☐
The cell is growing	☐	☐
The cytoplasm divides	☐	☐
Proteins for division are made	☐	☐
DNA is separated and moves to the two poles of the cell	☐	☐

3 Number these steps to put them in the correct order for the cell cycle. Then, work out which phase each step happens in.

ORDER	DESCRIPTION OF PHASE	PHASE
	The cell makes proteins needed for division, including centrosomes	
	The DNA in the nucleus is replicated	
	The cell divides down the middle, splitting into two daughter cells	
	The cell grows and makes new organelles like mitochondria and ribosomes	

4 Unscramble the anagram to reveal a keyword. Then write a definition for it in your own words!

HEARTSPINE = ______________________________

____

SOTISIM = ______________________________

____

TRAILPEEC = ______________________________

____

MONSTERCOE = ______________________________

____

© Emma The Teachie Limited 2025-Present. All rights reserved.

EXTEND → **Let's go deeper on this topic and extend your thinking — like a scientist!**

There are checkpoints that regulate progression through the cell cycle. For example, if a cell has damaged DNA, it will not enter the S phase. Or if the replicated DNA has errors, the cell will not progress to mitosis. Disruptions like this may result in programmed cell death, which is known as "apoptosis" (try saying it out loud, it's a great word!). Destroying a cell with damaged DNA benefits the organism, as that cell could otherwise become cancerous.

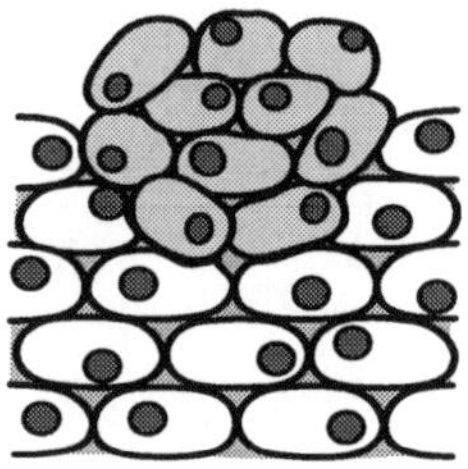

Skin tissue with cancerous cells forming a tumor

Occasionally, DNA errors are missed — if the mutation occurs in a gene that regulates the cell cycle, the cell can become cancerous. This means it divides uncontrollably, so that many more cells are produced than are needed. This is what causes tumors.

1. When errors are detected in DNA replication, what happens?

2. How are cancer cells different to normal body cells?

3. Chemotherapy is a type of cancer treatment. It can result in hair loss due to damage to hair follicles — these are the cells that make hair. Suggest how chemotherapy works and why it often causes hair loss.

REVIEW → **Check & strengthen your understanding by working through this activity.**

Set a 1 minute timer and try to summarize out loud everything you learned about the cell cycle inside one minute! Then, look back at the diagrams you made. Which bits did you forget to mention? Make a note of these in the space below.

© Emma The Teachie Limited 2025-Present. All rights reserved.

QUIZ ➜ **Show what you know by answering each of the questions.**

Put your noggin to the test!

1 Identify each phase of the cell cycle shown in the images.

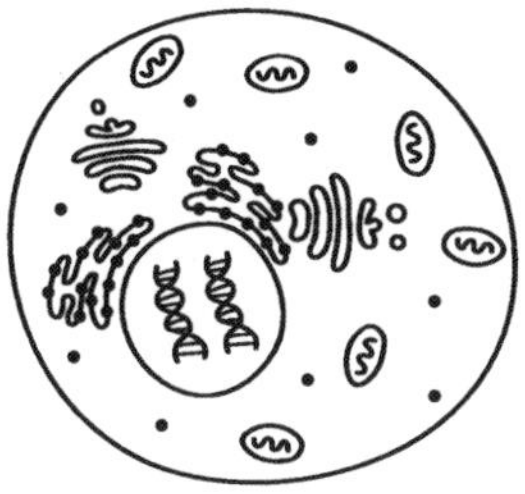

A

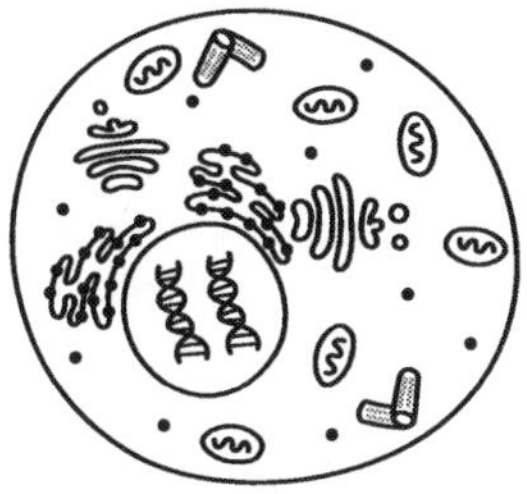

B

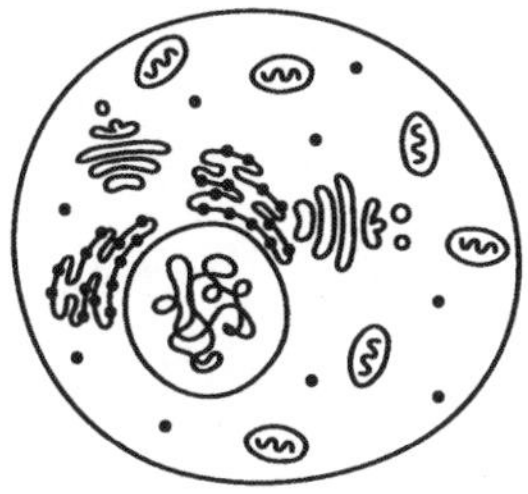

C

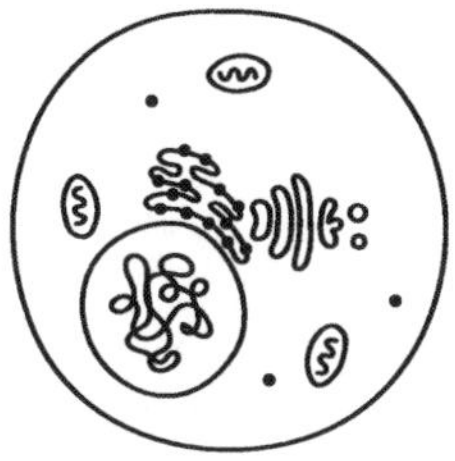

D

2 Which phase of the cell cycle are cells found in the majority of the time?

A. Mitosis

B. Meiosis

C. Interphase

D. G0 phase

3 Which stage of interphase follows cytokinesis?

A. Gap phase 1 (G1)

B. Gap phase 2 (G2)

C. Synthesis (S)

D. Meiosis

4 What occurs during synthesis?

A. Degradation of all organelles

B. DNA replication

C. Production of new organelles

D. Protein synthesis

5 What is the result of mitosis?

6 What happens during the G1 phase in the cell cycle?

7 Why do cells need to copy their DNA before division?

© Emma The Teachie Limited 2025-Present. All rights reserved.

Mitosis

PROCESS ➜ **Read the text passage below, then complete these tasks to process the info.**

HIGHLIGHT NEW KEYWORDS — SQUIGGLY UNDERLINE OLD KEYWORDS

UNDERLINE ANY DEFINITIONS — ADD ANY RELEVANT DOODLES IN SPACES

Mitosis is an important part of the cell cycle. It is how eukaryotic cells divide, forming new cells. New cells are important for a number of reasons.

Firstly, they allow the organism to grow. Secondly, new cells are used to repair any damage to tissues. And thirdly, some organisms reproduce asexually (there is no fusion of gametes), which requires new cells to be produced through mitosis.

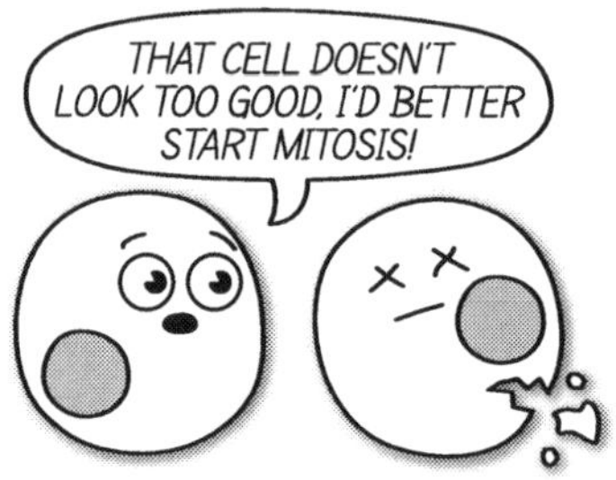

Mitosis alternates with interphase in the cell cycle. Remember that during interphase, the cell's DNA replicated, forming two copies of it. During mitosis, the DNA is separated so that each daughter cell ends up with one copy. The daughter cells are genetically identical to each other and to the original parent cell.

The Four Stages of Mitosis:

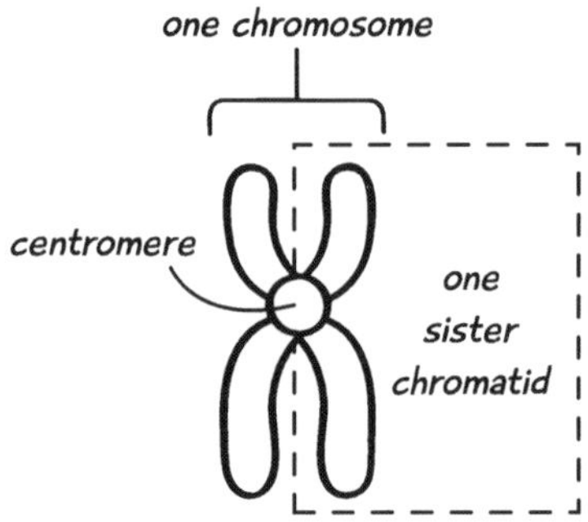

PROPHASE: Inside the nucleus, the DNA condenses into chromosomes. Each chromosome is made up of two sister chromatids. There are two because the DNA was replicated during interphase. They are joined at the centromere.

The nuclear envelope starts to break down. The two centrosomes start to move to opposite ends of the cell, which are known as poles. Spindle fibers start to extend between them.

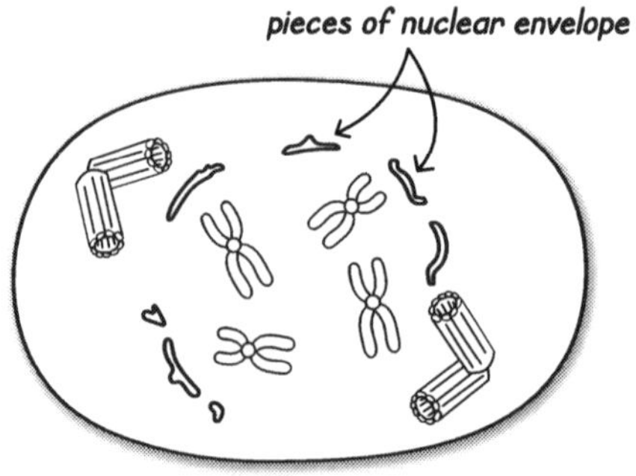

© Emma The Teachie Limited 2025-Present. All rights reserved.

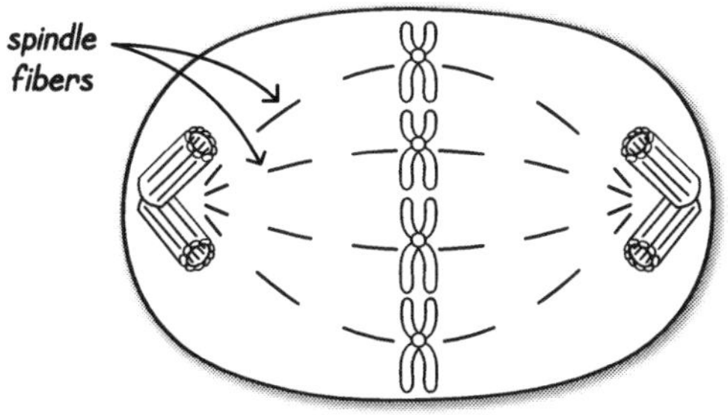

METAPHASE: The chromosomes are now free in the cytoplasm. Each centrosome has fully formed spindle fibers, collectively known as the spindle. The chromosomes line up along the middle of the cell and attach to the spindle by their centromeres.

TOP TIP: To remember this, think of the "M" for "Metaphase" as standing for "Middle" — the chromosomes line up in the middle of the cell!

ANAPHASE: The spindle fibers contract, separating the sister chromatids. Once they are separated, we refer to them as chromosomes again. They are pulled by their centromeres to poles of the cells. As they are pulled along by their centromeres, they appear V-shaped.

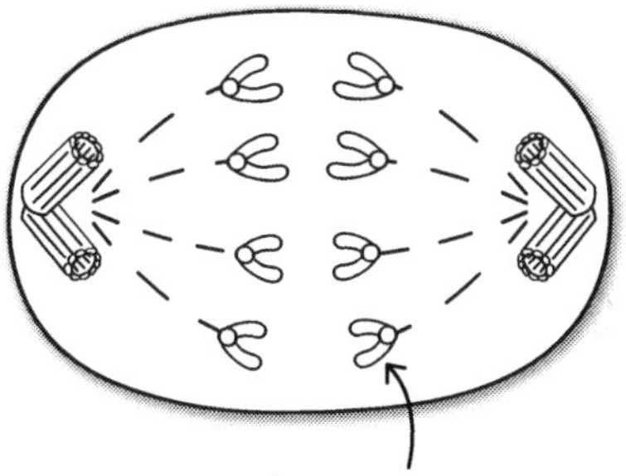

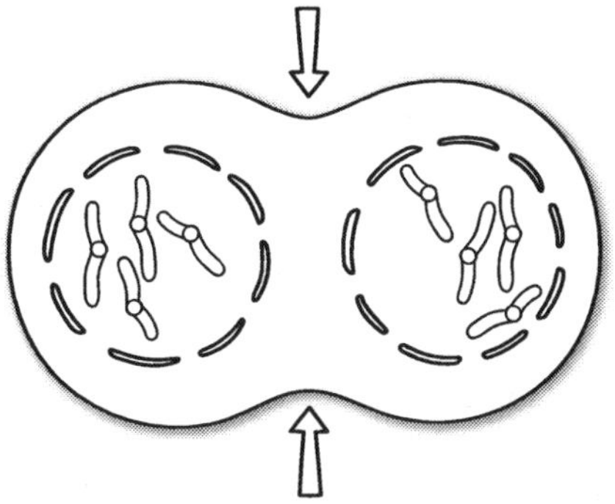

The cell membrane starts to pinch in, dividing the cytoplasm.

TELOPHASE: When the chromosomes reach the poles of the cell, they begin to uncoil, becoming longer and thinner. A nuclear envelope will form around each group of chromosomes. The spindle fibers break down.

The cytoplasm will start to divide in the process of **cytokinesis**.

When the cytoplasm has fully divided, cytokinesis is complete. The chromosomes are no longer visible. The daughter cells are genetically identical to each other and to the original parent cell. They are in interphase and will prepare for the next round of mitosis.

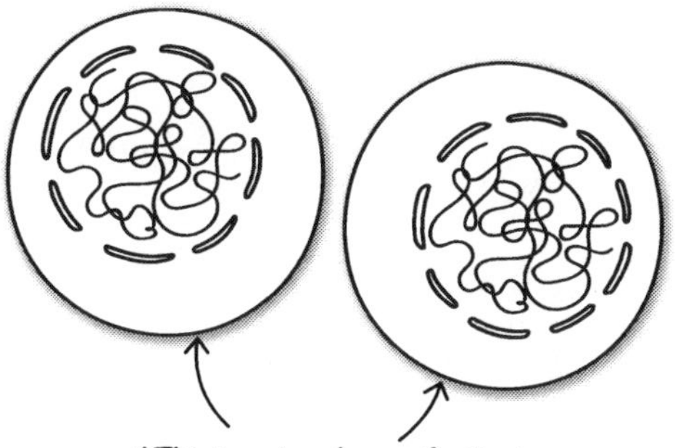

**This is not a phase of mitosis, it is the outcome: two daughter cells.*

© Emma The Teachie Limited 2025-Present. All rights reserved.

ACTIVATE → **Complete the activities to apply what you've learned from the text passage.**

1 Which of these are reasons for mitosis to occur? Color in the correct answers.

- REPAIR
- FORMATION OF GAMETES
- ASEXUAL REPRODUCTION
- APOPTOSIS
- GROWTH
- ENERGY PRODUCTION

2 It's important to remember the order of the phases of mitosis. We can do this by coming up with a mnemonic. This is a silly sentence or phrase in which the words start with the same letters as the words you are trying to remember!

Have a go at making your own mitosis mnemonic.

Example: ***P****arrots* ***M****eet* ***A****t* ***T****rees.*

P

M

A

T

Add your own doodle for your mnemonic if you want to!

3 Chromosome or chromatid? Look at each picture and decide if it shows a chromosome or a chromatid.

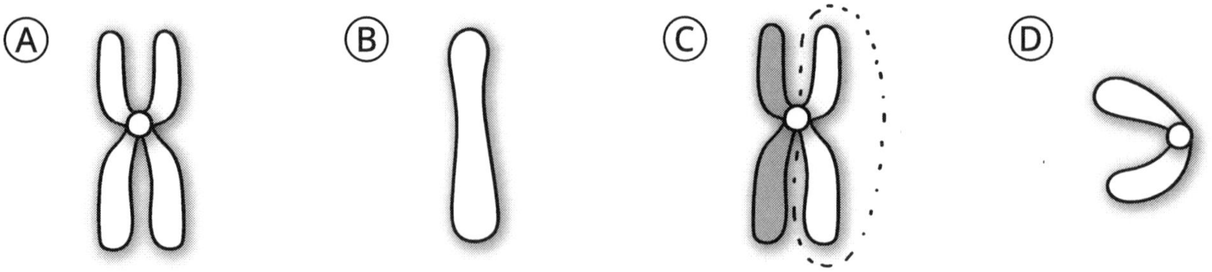

© Emma The Teachie Limited 2025-Present. All rights reserved.

4 Draw the stages of mitosis to help you learn what exactly is happening in each one.

- Leave out any organelles not mentioned in the text passage (i.e., no need to draw mitochondria, etc.).
- Clearly show what is happening to the nucleus and genetic material.
- Add labels for your diagrams — as many as possible to include all the details!

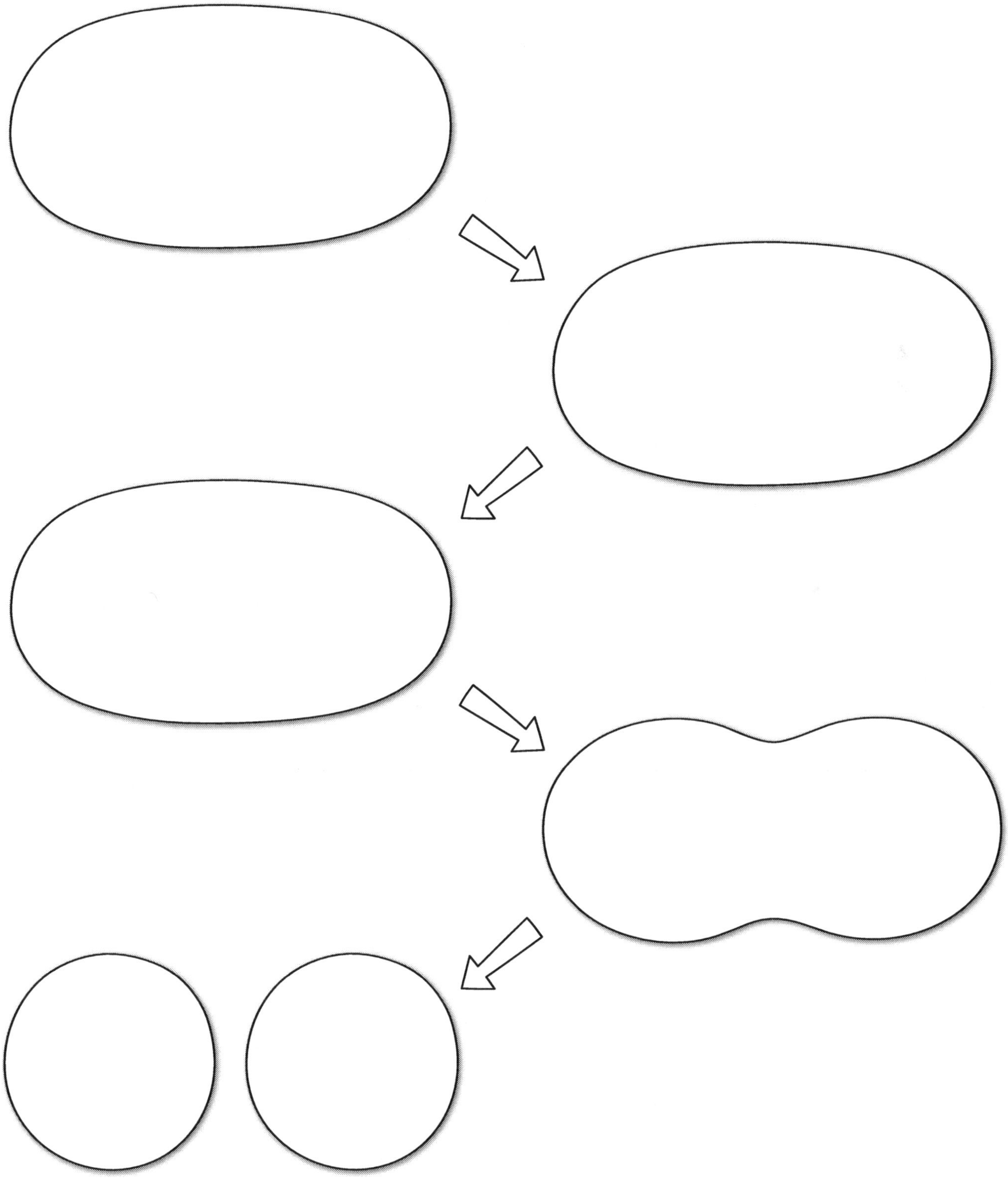

© Emma The Teachie Limited 2025-Present. All rights reserved.

EXTEND → **Let's go deeper on this topic and extend your thinking — like a scientist!**

The mitotic index of a tissue is the proportion of cells undergoing mitosis at once.
It can be calculated using this formula:

$$\text{Mitotic index} = \frac{\text{number of cells with visible chromosomes}}{\text{total number of cells in the field of view}}$$

← i.e., the number of cells undergoing mitosis

Create a key for the cells in this diagram. Then, calculate the mitotic index using the formula.

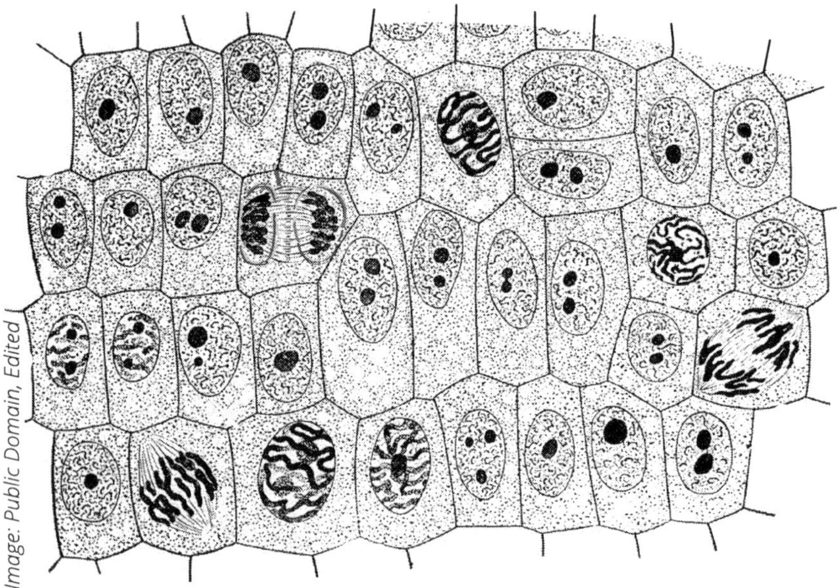

Image: Public Domain, Edited

KEY

☐ = Cells with visible chromosomes

☐ = Cells with no visible chromosomes

MITOTIC INDEX

Mitotic index = ______

=

TAKE IT FURTHER: For the cells undergoing mitosis, label them with the phase they are currently in.

REVIEW → **Check & strengthen your understanding by working through this activity.**

Which phase of mitosis does each of these events happen during?

1. ______ Chromosomes line up in the middle of the cell.
2. ______ The nuclear envelope reforms.
3. ______ DNA in the nucleus condenses into chromosomes.
4. ______ The spindle fibers begin to contract, pulling the sister chromatids apart.
5. ______ Chromosomes attach to the spindle fiber by their centromere.
6. ______ The chromosomes are pulled to opposite ends of the cell.
7. ______ A centrosome moves to each pole of the cell.

© Emma The Teachie Limited 2025-Present. All rights reserved.

QUIZ Show what you know by answering each of the questions.

Put your noggin to the test!

1 Identify each phase of mitosis shown in the images.

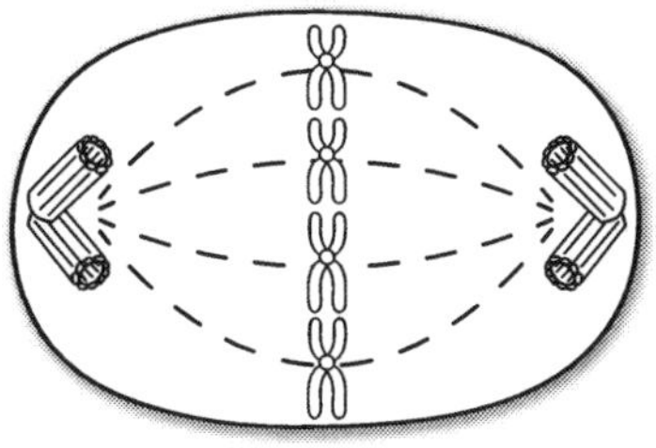

A

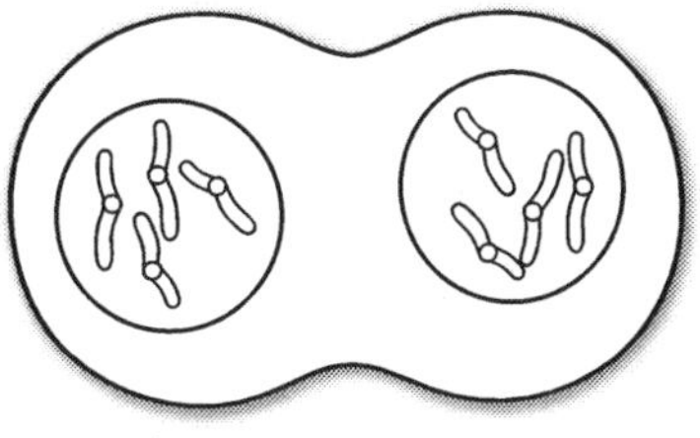

B

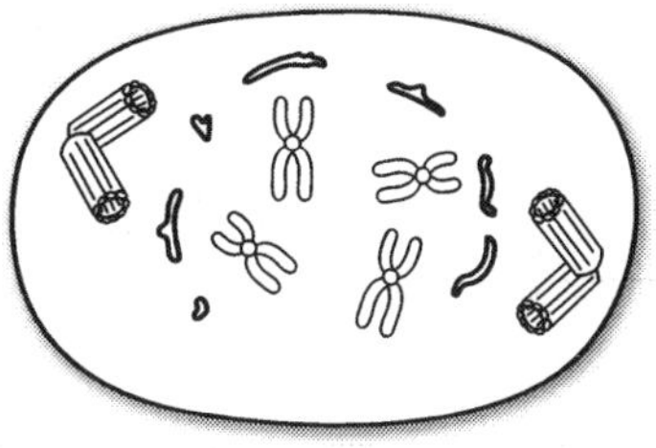

C

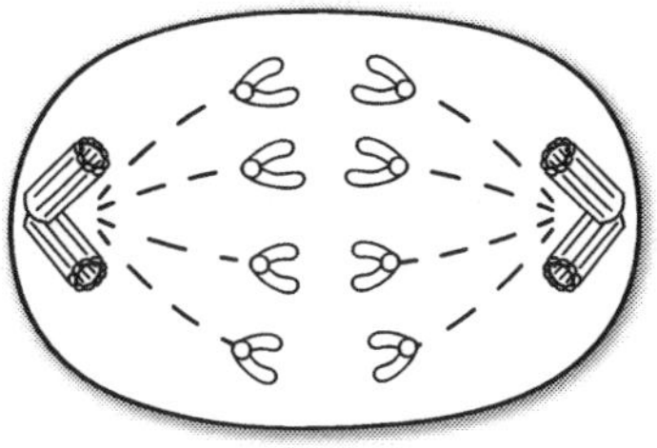

D

2 In which phase does the nuclear envelope reform?

A. Anaphase

B. Metaphase

C. Telophase

D. Prophase

3 What happens to DNA during prophase?

A. It decondenses

B. It replicates

C. It gets broken down with the nuclear envelope

D. It condenses

4 Which structure pulls chromosomes to opposite poles of the cell?

A. Spindle

B. Centrosome

C. Centrioles

D. Nuclear envelope

5 Write the stages of mitosis in order from first to last.

6 What is cytokinesis, and when does it happen?

7 Why does mitosis occur? Give three reasons.

© Emma The Teachie Limited 2025-Present. All rights reserved.

Meiosis

PROCESS → **Read the text passage below, then complete these tasks to process the info.**

HIGHLIGHT NEW KEYWORDS — SQUIGGLY UNDERLINE OLD KEYWORDS

UNDERLINE ANY DEFINITIONS — ADD ANY RELEVANT DOODLES IN SPACES

Meiosis is the special type of cell division that makes gametes. In animals, these are the sperm and egg cells. Females produce eggs and males produce sperm.

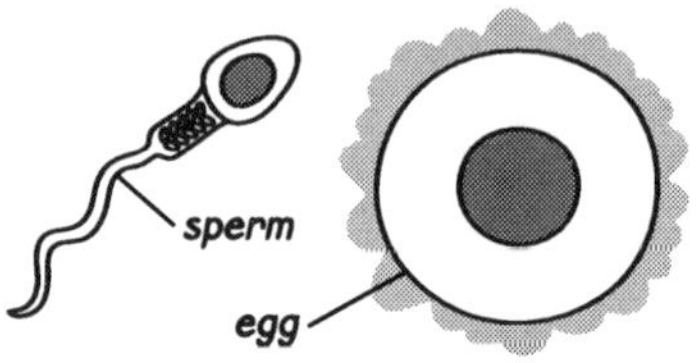

*Egg cells are ~1000x bigger than sperm!

Gametes only have one set of chromosomes. They are known as "haploid." Human gametes have 23 chromosomes.

When the egg and sperm fuse in the process of fertilization, they make a fertilized egg cell with two sets of chromosomes. We say that this cell is "diploid." It will divide many times to eventually produce all of the body cells. Our body cells have 46 chromosomes.

Meiosis happens after interphase, so the DNA has already replicated in preparation. Meiosis has **two** rounds of division, called Meiosis I and Meiosis II.

Meiosis I

We will use simplified diagrams with just four chromosomes!

Meiosis I starts with **prophase I**, when the nuclear envelope breaks down and DNA condenses into chromosomes. The chromosomes pair up with their matching partners, known as homologous chromosomes. Prophase I is also when crossing over happens, where parts of homologous chromosomes swap with each other, creating new genetic combinations.

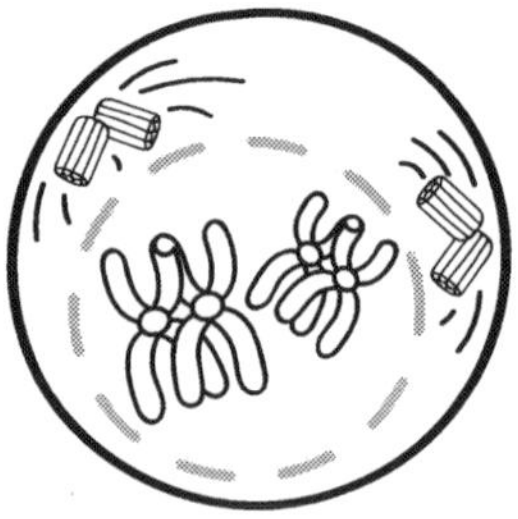

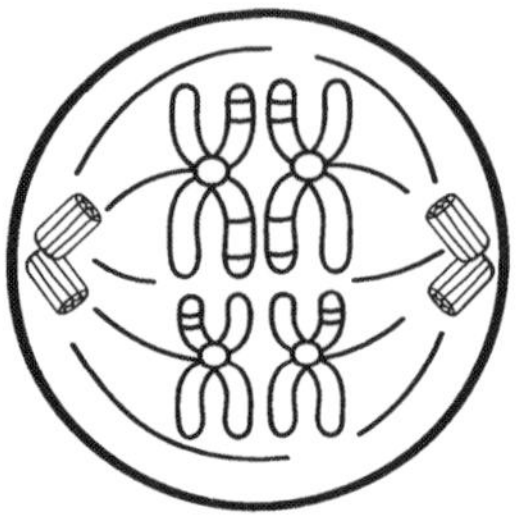

In **metaphase I**, the centrosome have reached the poles and the spindle fibers are fully formed. The chromosome pairs line up in the middle of the cell in random orientation. This random alignment leads to the gametes having different combinations of genes.

© Emma The Teachie Limited 2025-Present. All rights reserved.

Then, in **anaphase I**, the spindle fibers contract. The homologous chromosomes are pulled apart to opposite ends of the cell.

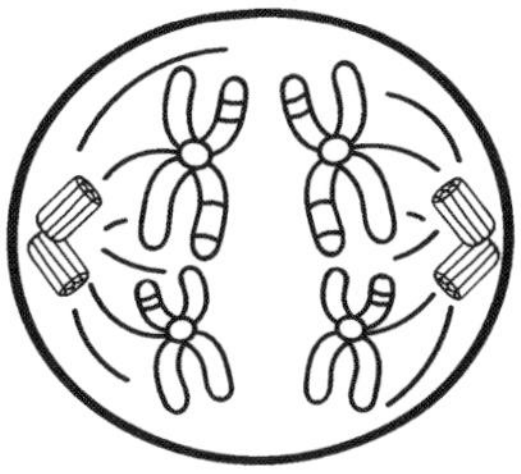

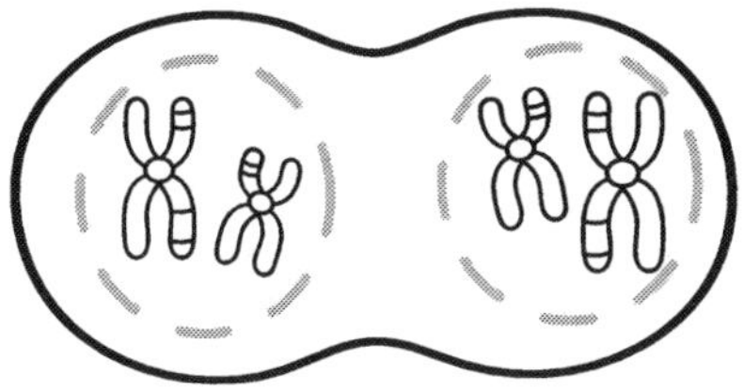

Finally, in **telophase I**, the nuclear envelope reforms at each pole, and then cytokinesis divides the cell into two haploid cells, each with 23 chromosomes. The chromosomes decondense.

Meiosis II

The two new haploid cells go through another round of division.

In **prophase II**, the chromosomes condense again. They are still formed of two sister chromatids joined at the centromere. Like before, the centrosomes move to the poles and spindle fibers begin to form.

In **metaphase II**, the chromosomes line up in the middle of the cell, attached to the spindle by their centromeres.

During **anaphase II**, the spindle fibers contract, separating the sister chromatids and pulling them to opposite ends of the cell.

Finally, in **telophase II**, a nuclear envelope reforms at each pole and the chromosomes decondense. Then, cytokinesis divides each cell into two. The result is four haploid gametes, each with 23 chromosomes (in humans). They are genetically **unique**.

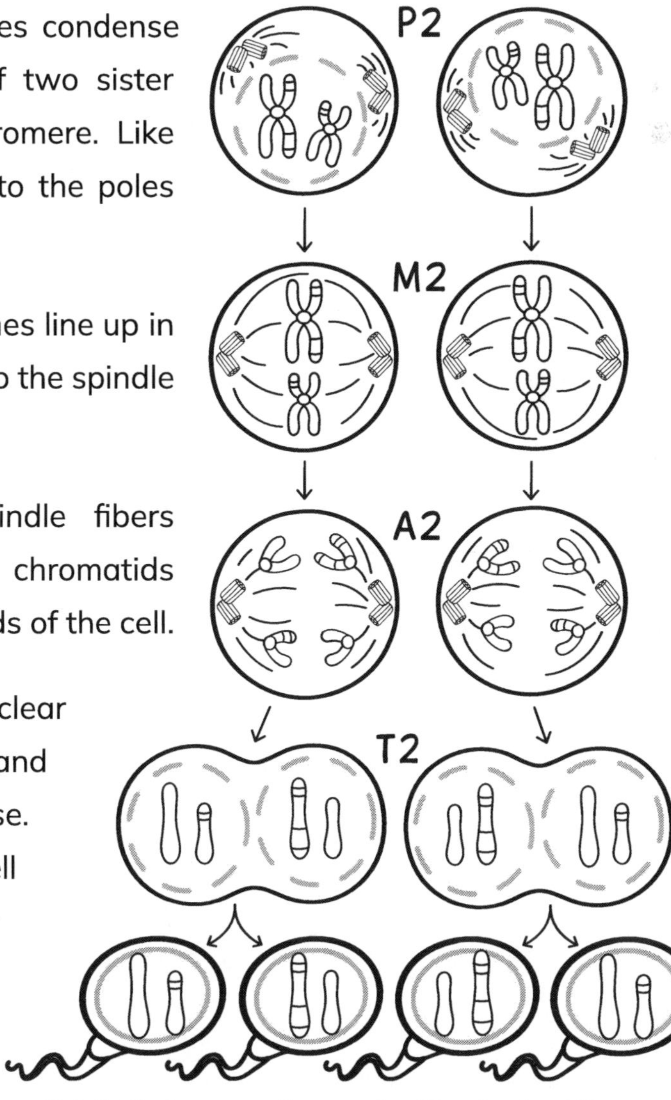

Note that the chromosomes in the sperm cells would have decondensed. They are drawn like this to illustrate that each daughter cell is genetically different.

© Emma The Teachie Limited 2025-Present. All rights reserved.

ACTIVATE → **Complete the activities to apply what you've learned from the text passage.**

1 There are a lot of keywords in this text passage! Let's make sure you know what they are. Match up the keyword to its correct definition.

Haploid	1. Division of the cytoplasm to form new cells
Diploid	2. Matching pairs of chromosomes, one from each parent
Crossing over	3. Cells with one set of chromosomes (23 in humans)
Cytokinesis	4. Sex cells, such as sperm and egg
Gametes	5. Cells with two sets of chromosomes (46 in humans)
Homologous chromosomes	6. Exchange of chromosome segments during Prophase I

2 A little bit more detail on homologous chromosomes. As you know, when a human egg and sperm fuse, they each contribute 23 chromosomes. The resulting body cell contains 46 chromosomes. Each of these chromosomes contains genes (lengths of DNA) for different characteristics, e.g., eye color, blood type. Homologous chromosomes are matching pairs — one from the egg and one from the sperm. They have the same type of genes in them.

Use the diagrams in the text passage on the previous pages to answer these questions:

A. How many **pairs** of homologous chromosomes are there at the start of meiosis? (In the diagram)

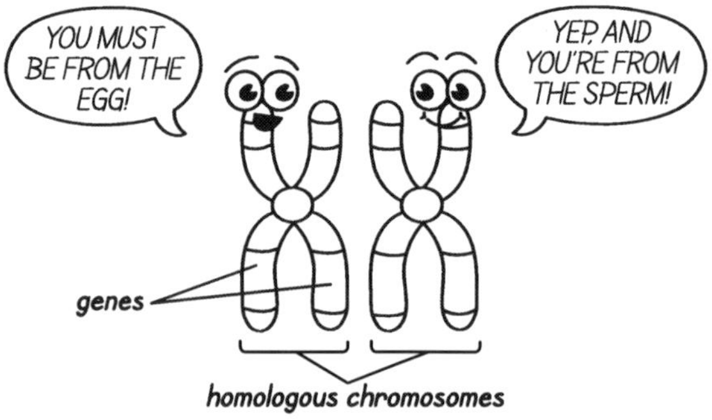

B. Which process happens between homologous pairs of chromosomes in prophase I?

C. During which phase are homologous pairs of chromosomes separated?

D. The diagram on the right shows the 46 chromosomes found in a human body cell. Suggest why they are **not** made up of two sister chromatids.

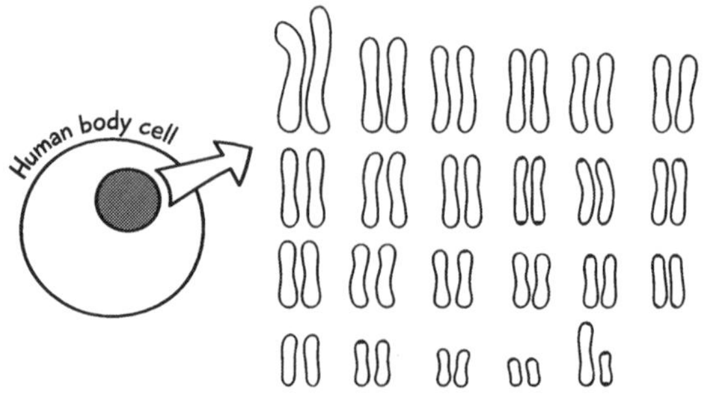

© Emma The Teachie Limited 2025-Present. All rights reserved.

3 These sentences about meiosis are unfinished. Complete them with the correct word/ phrase.

- Meiosis is a type of...
- Meiosis produces cells called...
- Before meiosis, the cell is in interphase. During this, the DNA is...
- At the start of meiosis I and II, the nuclear envelope needs to...
- Gametes have only one set of chromosomes, so we say they are...
- Gametes are genetically...
- Two gametes fuse together in the process of...
- Human body cells have two sets of...
- Human body cells can be described as being...

4 Now that you know about meiosis and mitosis, let's compare these two types of cell division! Sort these statements into the right part of the Venn diagram.

1. Daughter cells are genetically unique
2. Produces diploid cells
3. Two rounds of cell division
4. Occurs after DNA replication
5. Involves prophase, metaphase, anaphase, & telophase
6. Creates new sex cells
7. Daughter cells are genetically identical
8. A type of cell division
9. Produces 4 daughter cells
10. Produces haploid cells
11. One round of cell division
12. Creates body cells
13. Produces 2 daughter cells

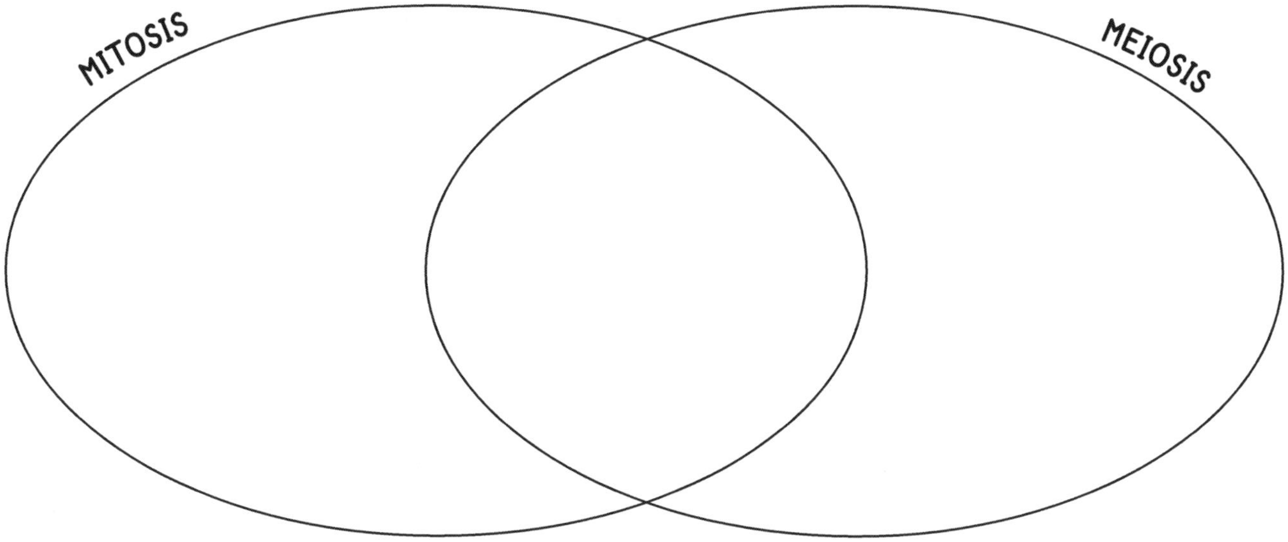

© Emma The Teachie Limited 2025-Present. All rights reserved.

EXTEND → **Let's go deeper on this topic and extend your thinking — like a scientist!**

We've discussed the number of chromosomes in human body cells and human gametes. But what about other animals? Use the numbers given to work out the missing chromosomes in the table.

Animal	Number of chromosomes	
	Body cell (2n)	Gametes (n)
Common fruit fly		4
Horse	64	
Dolphin		22
Lion	38	
Guinea pig		32
Atlas blue butterfly	458	
Dog		39

Did any of these surprise you? ______________________________

What can chromosome number tell you about the complexity of an organism?

____

REVIEW → **Check & strengthen your understanding by working through this activity.**

Decide if each statement is true or false. If you think it is false, explain why you think that.

STATEMENT	TRUE OR FALSE	If you circled false, why is it false?
1. Meiosis results in diploid cells.	T / F	______________
2. Crossing over happens during Prophase I.	T / F	______________
3. Human gametes contain 26 chromosomes.	T / F	______________
4. Meiosis produces four daughter cells.	T / F	______________
5. The purpose of meiosis is to make body cells.	T / F	______________

 © Emma The Teachie Limited 2025-Present. All rights reserved.

QUIZ ➔ **Show what you know by answering each of the questions.**

1 Identify each phase of meiosis shown in the images below.

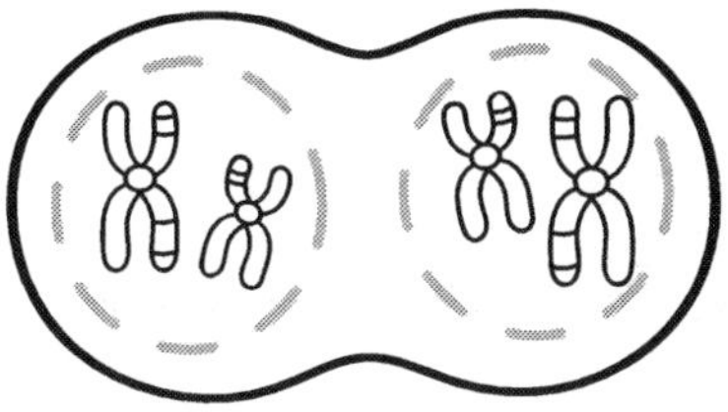

A

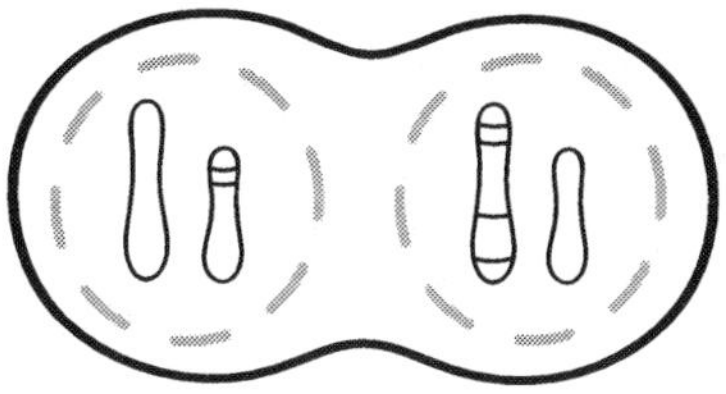

B

C

D

2 What type of cells are produced by meiosis?

A. Skin cells

B. Muscle cells

C. Neurons

D. Gametes

3 Crossing over occurs in which phase?

A. Metaphase I

B. Prophase II

C. Anaphase II

D. Prophase I

4 Which of the following best explains why meiosis is important?

A. It allows genetic variation in offspring

B. It repairs damaged tissues

C. It produces energy for cells

D. It makes identical copies of cells

5 How many daughter cells result from meiosis?

____

6 Define haploid and diploid.

____

____

7 What happens to homologous chromosomes during anaphase I?

____

____

____

© Emma The Teachie Limited 2025-Present. All rights reserved.

Sexual and asexual reproduction

PROCESS → **Read the text passage below, then complete these tasks to process the info.**

HIGHLIGHT NEW KEYWORDS — SQUIGGLY UNDERLINE OLD KEYWORDS

UNDERLINE ANY DEFINITIONS — ADD ANY RELEVANT DOODLES IN SPACES

Living things need to reproduce so their species can continue. There are two main types of reproduction: asexual reproduction and sexual reproduction. Each has its advantages and disadvantages, and different organisms use different methods.

Asexual Reproduction

In asexual reproduction, only one parent is needed. The offspring are genetically identical to the parent, which means they are clones.

Asexual reproduction is usually quick and allows organisms to reproduce rapidly when conditions are good. It allows isolated individuals to produce offspring on their own.

Lone female sharks in aquariums have been known to give birth to healthy pups!

However, since all the offspring are identical, they may not survive well if the environment changes. They are more likely to be affected by new diseases. If a harmful mutation occurs, it is always be passed onto future generations.

Methods of Asexual Reproduction

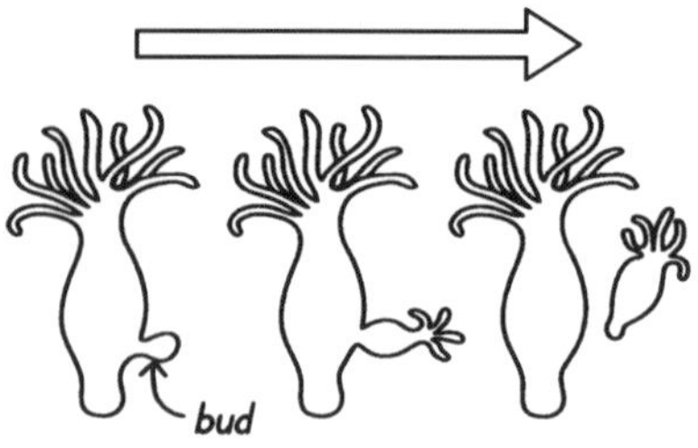

Budding: There is increased cell division at one site of an organism. It grows and eventually breaks off to become its own new individual. E.g., hydra, yeast.

Fragmentation: Part of a mature organism breaks off and develops on its own to become a new organism. E.g., fungi, star fish, sea sponge.

© Emma The Teachie Limited 2025-Present. All rights reserved.

Binary fission: This is how prokaryotes divide. The DNA is replicated, and the cell grows and elongates. DNA molecules move to opposite poles, and the cell divides in two. E.g., E. coli.

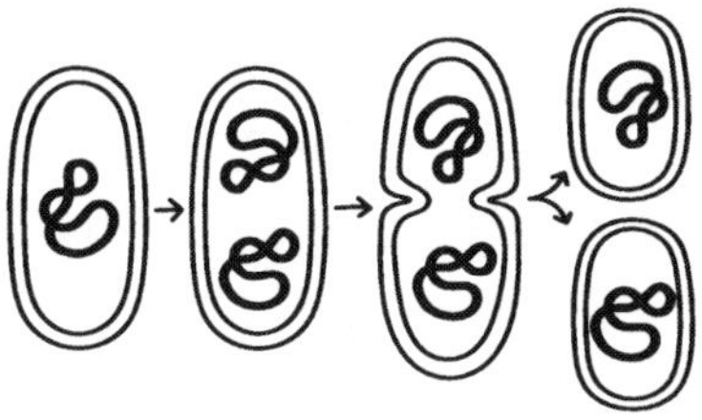

Parthenogenesis: An embryo develops from an unfertilized egg. The offspring may be full or half clones of the mother. E.g., Komodo dragons, drone bees.

Vegetative propagation: Plants produce new individuals directly from the parent organism. E.g., runners in strawberry plants.

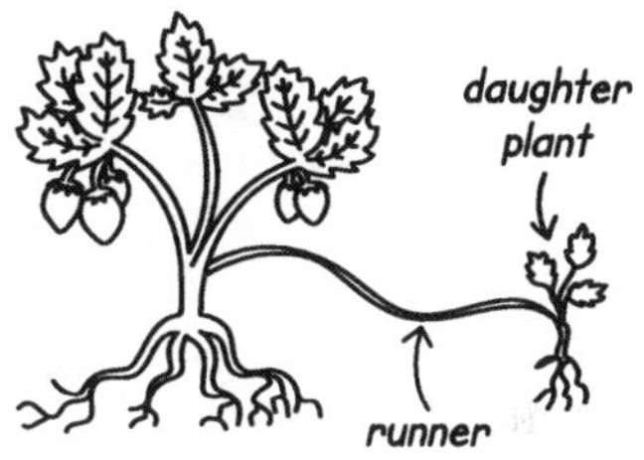

Sexual Reproduction

In sexual reproduction, two parents are involved. Each parent provides half of the genetic material through special cells called gametes. When two gametes combine, they form a fertilized egg with a unique mix of DNA. This means offspring are genetically different from their parents and from each other. Most animals and many plants use sexual reproduction.

Although it takes more time, energy, and usually produces fewer offspring, this method of reproduction increases genetic diversity. This helps species adapt to new environments. And if a new disease emerges, there may be some offspring who have immunity. Harmful mutations can also be removed from the population as they are not always passed on.

Certain plants and simple animals can switch between sexual and asexual reproduction, depending on the conditions surrounding them. When conditions are good, asexual reproduction may allow them to multiply quickly. When times are tough, sexual reproduction creates variety, giving some offspring a better chance of survival.

© Emma The Teachie Limited 2025-Present. All rights reserved.

ACTIVATE → **Complete the activities to apply what you've learned from the text passage.**

1 Let's lock down those keywords. Write a definition for sexual and asexual reproduction. Give an example of an organism that uses each method to reproduce.

WHAT IS SEXUAL REPRODUCTION:

WHAT IS ASEXUAL REPRODUCTION:

2 Complete the table to show the advantages and disadvantages of each type of reproduction.

	Sexual Reproduction	Asexual Reproduction
Advantages		
Disadvantages		

© Emma The Teachie Limited 2025-Present. All rights reserved.

3

Match up the method of asexual reproduction to its description and example!

METHOD	DESCRIPTION	EXAMPLE
Parthenogenesis	Plants produce new offspring directly from the parent organism.	Listeria monocytogenes
Fragmentation	A new organism develops from a small outgrowth on the parent organism that eventually breaks off.	Spider plant runner
Budding	A prokaryote replicates its DNA, grows, and divides into two cells.	Yeast
Vegetative propagation	An embryo develops from an unfertilized egg. No sperm are involved.	Sea sponge
Binary fission	Part of a mature organism breaks off and develops on its own to become a new organism.	Water fleas

4

Sexual reproduction produces genetic variation in offspring. Let's look at this in a beetle population.

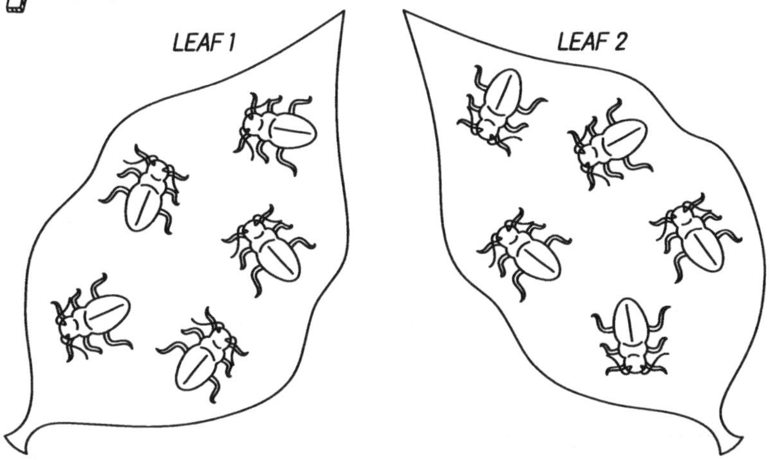

Genetic variation causes some beetles to be green and some beetles to be brown.

On each leaf, color some beetles green and some of them brown.

Leaf 1 is grown during a normal year. Color the leaf in green.

Leaf 2 is grown during a drought year. Color the leaf in brown.

A. Beetles are eaten by other insects, birds, small mammals, etc. Draw an "X" over any beetles that you think would not survive on each leaf. Then answer: why would those beetles not survive?

B. Now image the beetles are all offspring from a green beetle producing asexually. What color would they be? How many beetles would you expect to survive on leaf 1 and on leaf 2?

C. Use this scenario to explain the main benefit of sexual reproduction over asexual reproduction.

© Emma The Teachie Limited 2025-Present. All rights reserved.

EXTEND → **Let's go deeper on this topic and extend your thinking — like a scientist!**

There have been cases where sharks and stingrays who have been kept alone or in all female aquariums have had offspring! This only happens after years, under extreme conditions.

Design a new organism that uses both sexual and asexual reproduction. Draw it and describe:

- What environment it lives in
- When it uses asexual reproduction
- When it switches to sexual reproduction

REVIEW → **Check & strengthen your understanding by working through this activity.**

A student has written notes about sexual and asexual reproduction. Unfortunately, they have made a few mistakes! You are going to help them - find and underline their mistakes, then rewrite their summary with corrections.

Sexual reproduction involves two parents. It is when two body cells fuse together, combining the RNA. This creates genetic speciation in the offspring.

Asexual reproduction involves one or two parents. The offspring are genetic twins of the parent.

Sexual reproduction is faster than asexual reproduction, and the variation in offspring gives them a greater chance of surviving in a new environment. Asexual reproduction is slower but allows more offspring to be produced if the conditions are favorable.

© Emma The Teachie Limited 2025-Present. All rights reserved.

QUIZ → **Show what you know by answering each of the questions.**

Put your noggin to the test!

1 Identify each method of asexual reproduction shown in the images.

A

B

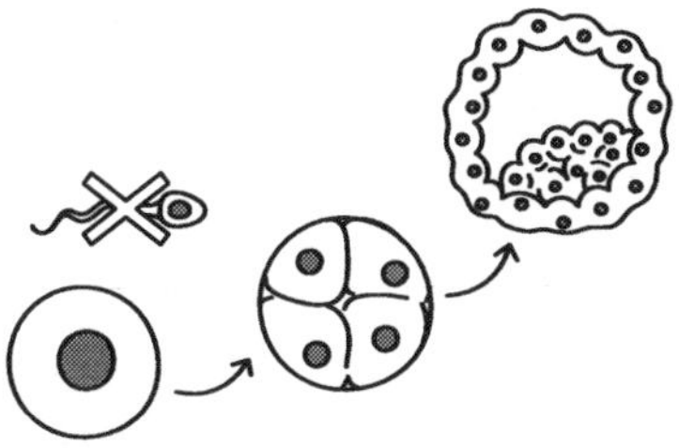

C

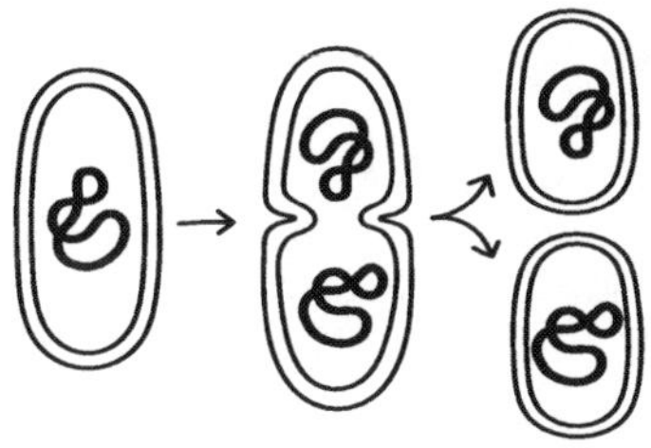

D

2 Which type of reproduction requires gametes?

A. Sexual reproduction

B. Asexual reproduction

C. Binary fission

D. Parthenogenesis

3 Which is a disadvantage of asexual reproduction?

A. It is slow

B. Offspring are genetically identical

C. It requires two parents

D. It increases diversity

4 What is formed when two gametes combine?

A. A clone

B. A fertilized egg

C. A haploid cell

D. A bud

5 Give one advantage of asexual reproduction.

____

6 Define asexual reproduction.

____

____

7 Why is sexual reproduction beneficial for species survival?

____

____

____

© Emma The Teachie Limited 2025-Present. All rights reserved.

UNIT 4 REVIEW CROSSWORD

ACROSS

5 The genetically identical offspring made by asexual reproduction
7 Offspring with different traits due to sexual reproduction are said to be...
9 Stage of the cell cycle when a cell grows and prepares to divide
11 Type of cell division that produces two identical body cells
12 Simple cell division used by bacteria to reproduce
14 Asexual reproduction where a new organism grows from the parent's body
15 The process of making an exact copy of DNA
16 Having half the usual number of chromosomes
17 Sex cells like sperm and eggs
18 Division of the cytoplasm to form two new cells
20 Tightly coiled DNA, visible during cell division
21 Exchange of chromosome pieces during meiosis that creates variation in gametes

DOWN

1 First stage of mitosis, when chromosomes become visible
2 Type of cell division that makes sex cells
3 Stage of mitosis when sister chromatids are pulled apart to opposite poles of the cell
4 The part of interphase when DNA is copied
6 The process of fixing damaged cells or tissues; one reason for mitosis to occur
8 Joining of sperm and egg to form a new cell
10 Structures that help move and separate chromosomes in mitosis and meiosis
11 Stage of mitosis and meiosis when chromosomes line up in the center of the cell
13 Phase where nuclear membranes reform around separated chromosomes
19 Male gamete used in sexual reproduction

© Emma The Teachie Limited 2025-Present. All rights reserved.

REMEMBER: ONLY USE THE ANSWER KEYS ONCE YOU HAVE DONE THE ACTIVITY/QUIZ YOURSELF! (EVEN IF SOMETIMES YOU HAVE TO MAKE AN EDUCATED GUESS!)

ANSWER KEYS

FILL THE SPACE WITH IMPORTANT INFO! MAYBE THERE IS SOMETHING YOU GOT WRONG OR A KEYWORD YOU STRUGGLE TO REMEMBER THE MEANING OF!

© Emma The Teachie Limited 2025-Present. All rights reserved.

UNIT 1 Review Crossword *p. 26*

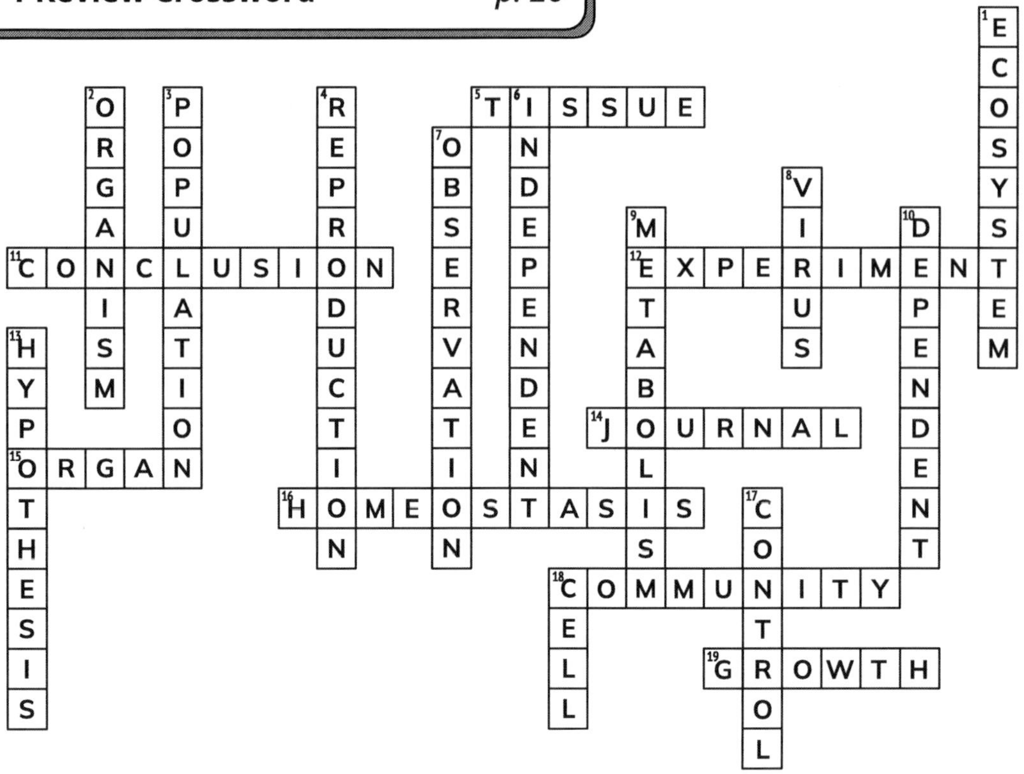

UNIT 2 Review Crossword *p. 54*

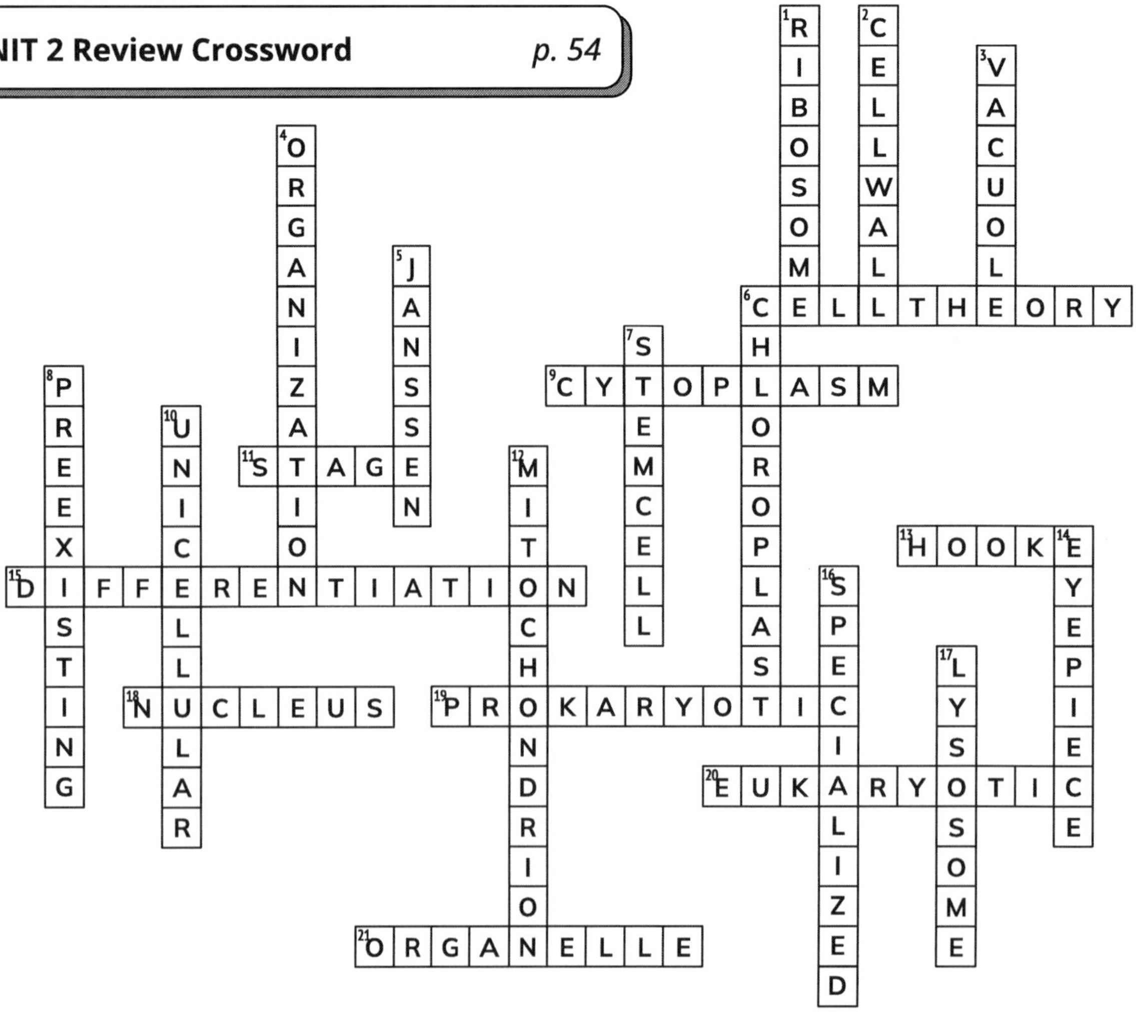

© Emma The Teachie Limited 2025-Present. All rights reserved.

UNIT 3 Review Crossword

p. 86

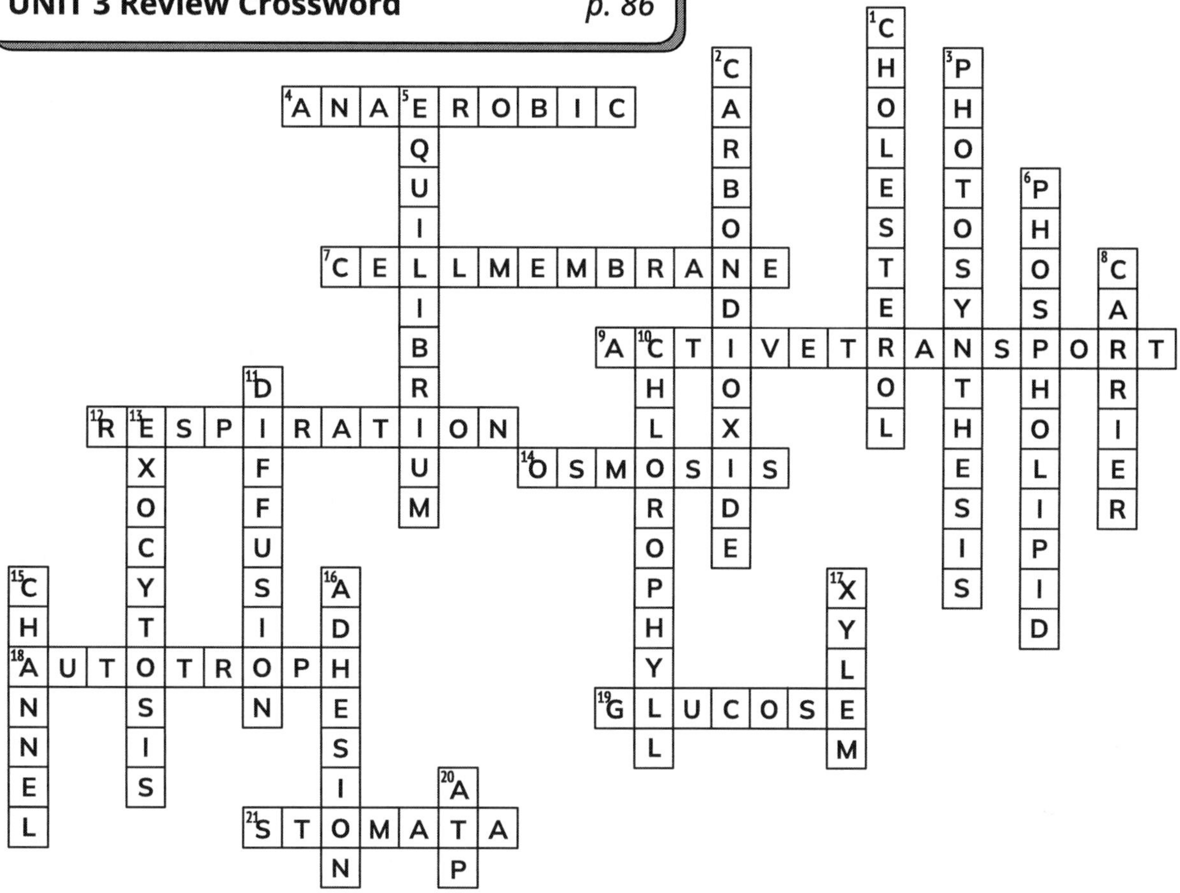

UNIT 4 Review Crossword

p. 112

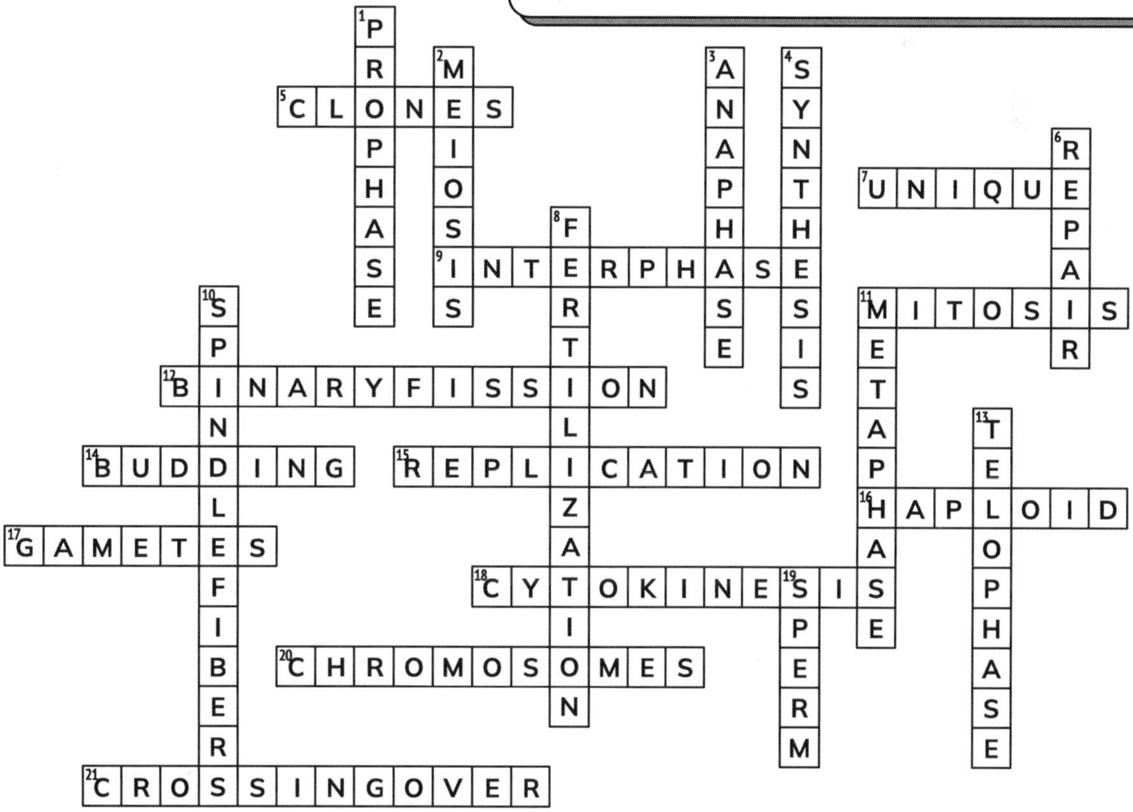

© Emma The Teachie Limited 2025-Present. All rights reserved.

ACTIVATE

1. The lines should match up as follows: Experiment - E, Hypothesis - A, Conclusion - D, Research - B, Report - H, Analysis - F, Question - G, Observation - C.
   The numbers should read (from top to bottom): 5, 4, 7, 3, 8, 6, 2, 1.

2. Josh: IV - material seed is planted in, DV - volume of water that drains through the soil, CV - length of time, volume of water poured into the pot, type of seed, amount of material.

   Alexis: IV - type of exercise, DV - heart rate, CV - length of exercise, length of rest between activities.

   Luca: IV - temperature of room, DV - number of cricket chirps, IV - length of time listening to chirps, adjustment time to new temperature, species/individual cricket used.

3. Alexis: As the intensity of exercise increases, heart rate also increases. When Alexis was walking, her heart rate was 85 bpm, during jogging it increased to 126 bpm, and while running it rose to 184 bpm. When you exercise more intensely, your muscles use more oxygen and produce more CO_2. The heart must pump faster to deliver more oxygen and remove CO_2.

   Luca: As temperature increases, the number of cricket chirps per minute also increases. At 15°C, the crickets chirped 20 times per minute; at 20°C, they chirped 28 times; and at 25°C, they chirped 39 times. Crickets are cold-blooded, meaning their body temperature depends on the environment. Warmer temperatures speed up their metabolism and muscle activity, causing them to chirp more frequently.

EXTEND

Open ended — your hypothesis should be written in the "if... then... because..." format.

Your experiment can be either diagrams to explain it, or a step by step written method. It should make it clear what is changed (independent variable), measured (dependent variable), and kept the same (control variables).

REVIEW

A. Senses
B. Testable
C. Answered
D. Hypothesis; then
E. Independent; dependent; control
F. Analyze; patterns
G. Conclusion; claim-evidence-reasoning (CER)
H. Reports; journals

QUIZ

1. A) dependent, B) control, C) independent
2. B
3. C
4. B
5. Something you notice with your senses
6. Claim - Evidence - Reasoning
7. If there is more light, then the plant will grow more because plants use light for photosynthesis. *You can also say "If the light intensity is higher, then..." — this is great scientific vocabulary!*

© Emma The Teachie Limited 2025-Present. All rights reserved.

Characteristics of Life *pp. 14–19*

ACTIVATE

1. Unicellular: bacterium, amoeba, paramecium. Multicellular: human, cat, tree.

2. In order (from top to bottom): growth, homeostasis, metabolism, response to stimuli, cells, evolution, reproduction: living things can reproduce to create new offspring.

3. Your explanation must include at least three characteristics of life.
   Sample answer: Robots are not alive because they are not made from **cells**; they are made from other materials like metal and plastic. They cannot **grow** like a living thing grows. And they do not have a **metabolism**, as chemical reactions are not taking place inside cells.

4. Rock: nonliving - it does not have any characteristics of life, e.g., no cells.
   Seaweed: living - it can grow, reproduce, respond to stimuli, and more.
   Snowflake: nonliving - although it can grow, it does not have cells, a metabolism, etc.
   Wooden table: nonliving - although it is made of cells, it cannot grow or evolve, etc.
   Grasshopper - living - it can reproduce, respond to stimuli, all characteristics of life.

5. I think zonkeys are... you can argue it either way. There is no right or wrong answer here.

   Sample answer for "alive" - they have almost all of the characteristics of life: They are made of cells, can grow, metabolize, respond to stimuli, and carry out homeostasis. Even though they cannot reproduce or evolve, they are still living creatures.

   Sample answer for "not alive" - even though they have many characteristics of life, they are unable to reproduce as a population, and therefore they cannot evolve over time. So there are two characteristics of life they do not have.

EXTEND

Arguments for: they are capable of evolving, and they can reproduce to make new viruses (although not on their own). And while they don't have cells, they do have genetic material like DNA or RNA.

Arguments against: they cannot reproduce on their own, they do not metabolize, they do not do homeostasis, unclear if they can respond to stimulus.

Explain what you think: share your own opinion. Scientists have a consensus that viruses are nonliving, but it is up to you to decide what you think.

REVIEW

All seven characteristics of life should be listed around the bubble with doodles to aid memory.

QUIZ

1. A) reproduction, B) growth, C) response to stimuli, D) evolution *(the stick insect has evolved to become camouflaged)*
2. A
3. B
4. B
5. C
6. Unicellular; multicellular.
7. Any one example, e.g., regulating temperature (sweating, shivering); regulating blood sugar (producing hormones insulin, glucagon); regulating water balance (kidney function); regulating blood pressure (heart rate and blood vessel contraction).

© Emma The Teachie Limited 2025-Present. All rights reserved.

Levels of Organization *pp. 20–25*

ACTIVATE

1. The lines should match up as follows: Organ - C, Tissue - E, Cell - A, Organ system - B, Organism - D.
   The numbers should read (from top to bottom): 3, 2, 1, 4, 5.

2. A) organ (lungs), B) organism (duck), C) cell (neuron), D) tissue (muscle tissue), E) organ system (circulatory system - spot the heart!), F) tissue (stomach epithelial tissue), G) population (fish), H) organ system (digestive system)

3. Open ended — you should have doodled a few different organs and their names. Some options: brain, heart, lungs, kidneys, liver, small intestines, gall bladder, etc.

4. Open ended — you should have picked an organ system and described the cells, tissues, and organs that are included in this system. (Note: You do not need to describe every single one of these, just a few examples.)

EXTEND

Open ended — you should have picked one organ or tissue that is damaged and described the impact of this, the body's response, and possible medical intervention.

Sample answer:

Damaged organ/tissue: heart.

Impact of damage on organ system & organism: cardiovascular system cannot pump blood as efficiently around the body. Leads to shortness of breath, fatigue, as not enough oxygen is getting to cells.

Body's response + possible medical intervention: depends on condition; e.g. heart can compensate (increasing heart rate, constricting blood vessels, etc.) Interventions could include lifestyle modifications to diet, exercise, smoking habits; medication like aspirin and beta-blockers; or procedures like stents or fitting a pacemaker.

REVIEW

1. Organism
2. Function
3. Tissue
4. Organ system
5. Organ
6. Cell

QUIZ

1. A) cell, B) organism, C) tissue, D) organ system (it is the urinary system - spot the kidneys and bladder!)
2. D
3. C
4. C
5. Organ
6. Living
7. False

© Emma The Teachie Limited 2025-Present. All rights reserved.

Cell Theory *pp. 28–33*

ACTIVATE

1. These should be in your own words. All organisms are made of cells. Cells are the basic unit of life. All cells come from other cells (preexisting cells).

2. Lines should be matched up in this order:

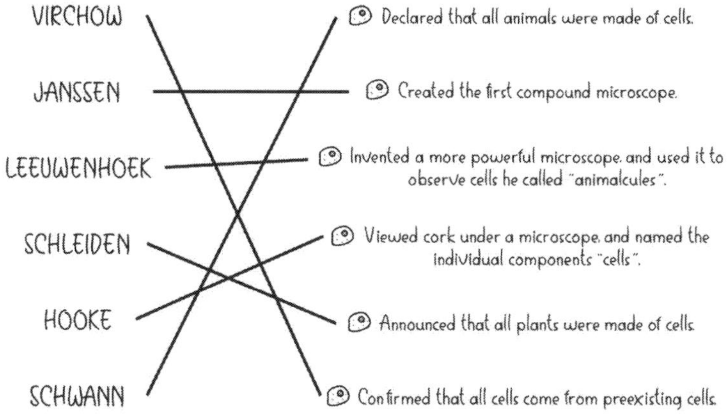

3. A) nosepiece, B) objective lens, C) stage, D) diaphragm, E) light source, F) eyepiece, G) arm, H) stage clip, I) coarse focus knob, J) fine focus knob, K) base. These should be color coded to match the piece of the microscope they are labeling.

EXTEND

1. Sterilization (Note: Disinfection is not the same — it does not kill all life)

2. The sterile broth didn't spoil even after weeks. This proves that no microorganisms were spontaneously generated in it. They have to come from existing microorganisms.

3. Open ended.

   Sample answer: diagram of three glass jars with a piece of meat in each. One jar has a lid on top, one jar has mesh over the top (attached with an elastic band), and one jar is open to the air. Jars should be left for over 24 hours (48 or 72 hours would be suitable). The expected results are that the jar with the lid has no maggots, the jar with the mesh would have maggots on top of the mesh only, and the jar with open meat has maggots on the meat. This proves maggots come from fly eggs, not the meat itself.

REVIEW

Missing words: preexisting; organization/life; cells.

QUIZ

1. A) true, B) false, C) false, D) false
2. D
3. B
4. B
5. It allowed scientists to view cells for the first time.
6. The stage
7. Coarse focus knob and fine focus knob

© Emma The Teachie Limited 2025-Present. All rights reserved.

Eukaryotic and Prokaryotic Cells pp. 34–39

ACTIVATE

1. Labels — first part of the task shown on the left. "Take it further" task shown on the right.

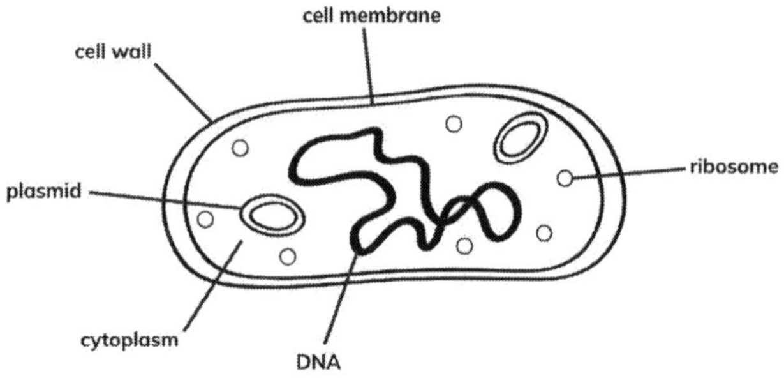

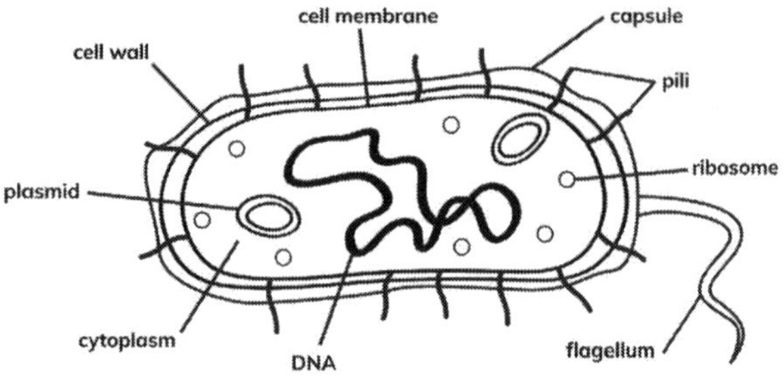

2. Left poster: Name of cell - prokaryotic; Defining features - cell wall, free DNA (not in a nucleus), possibly plasmids.
   Right poster: Name of cell - eukaryotic; Defining features - nucleus, mitochondria, ribosomes.

3. Eukaryotes: 1, 4, 8, 10, 12. Prokaryotes: 2, 6, 7, 9. Both: 3, 5, 11.

4. A) prokaryotic, B) eukaryotic, C) eukaryotic, D) prokaryotic, E) eukaryotic, F) eukaryotic (we can't see the cells to decide if they have a nucleus or not, but it is multicellular so it must be eukaryotic). G) and H) are open ended answers. Your diagrams should clearly show if there is a nucleus (eukaryotic), or if it has free DNA (prokaryotic), or if it is multicellular (eukaryotic).

EXTEND

Open ended — you should have your own suggestions of what the capsule does.

Some functions of the prokaryotic capsule (Note: You do not need to have all of these, and you may have some extra ones):

- Prevents drying out
- Shields against harmful chemicals
- Helps bacteria stick to surfaces
- Helps bacteria stick to each other
- Makes bacteria more harmful
- Protects bacteria from viral infections (yes, some viruses infect bacteria!)
- Helps bacteria hide from and evade the immune system

REVIEW

1. Yes | No
2. 10μm–100μm | 0.1μm–5.0μm
3. Yes | No
4. Nucleus | Free in cytoplasm
5. Yes | No
6. No | Yes
7. Mitosis | Binary fission

QUIZ

1. A) nucleus, B) plasmid, C) cell wall, D) flagellum
2. D

© Emma The Teachie Limited 2025-Present. All rights reserved.

3. C
4. B
5. Eukaryotic cells are cells that **have a nucleus**; Prokaryotic cells are cells that **do not have a nucleus**.
6. Binary fission
7. Eukaryotic

Cell Organelles pp. 40–47

ACTIVATE

1. Labels, and stars on the plant cell:

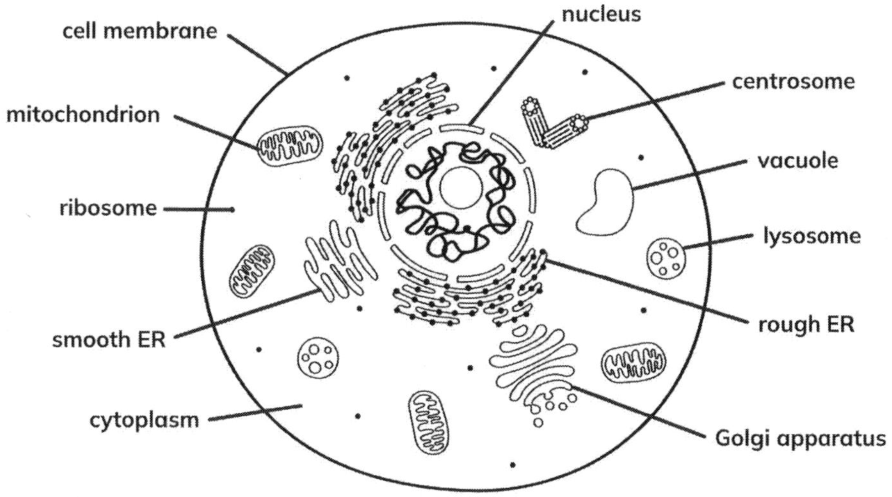

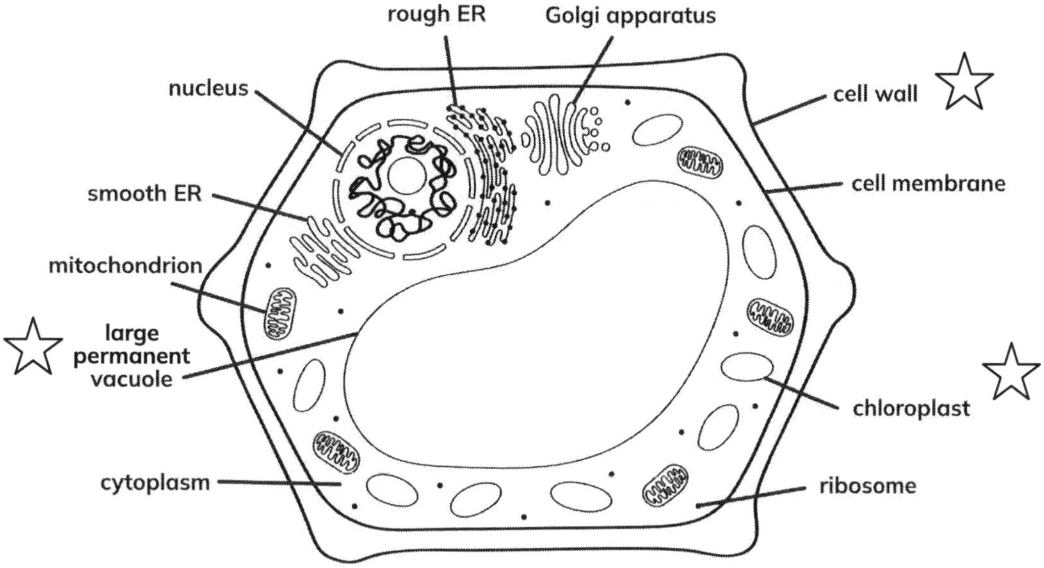

2. All keywords from the keyword box should be used. These are in bold below.

   Cell membrane: **controls** what enters and exits the cell.
   Cell wall: provides **strength** and structure to the cell.
   Nucleus: the **brain** of the cell, contains the cell's **DNA** and controls the cell's activities.
   Ribosome: produces **proteins** for the cell's own use.

© Emma The Teachie Limited 2025-Present. All rights reserved.

Smooth ER: produces carbohydrates, lipids, and hormones. Detoxifies drugs and poisons.
Rough ER: produces **proteins** for transport out of the cell.
Mitochondrion: the **powerhouse** of the cell, produces energy from cellular **respiration**.
Golgi apparatus: modifies, tags, and **packages proteins** and lipids.
Lysosome: breaks down and recycles **waste**, organelles, and cells using digestive enzymes.
Vacuole: **stores waste** products in animal cells. In plants, it also provides support, and recycles **waste**.
Chloroplast: carries out **photosynthesis** to make food for the plant.
Cytoplasm: contains all organelles, and helps transport substances.
Centrioles: form spindle fibers used in cell **division**.

3. Nucleus - both
   Cell wall - plant cell
   Mitochondria - both
   Cell membrane - both
   Chloroplasts - plant cell
   Lysosomes - animal cell
   Large central vacuole - plant cell
   Centrosomes - animal cell

EXTEND

1. Chloroplasts

2. Eye spot for detecting light; contractile vacuole for movement

3. A protist

REVIEW

1. False — Ribosomes can be free
2. True
3. False — Ribosomes produce proteins
4. False — The plant cell wall is like this
5. True
6. True
7. True
8. False — Animal cells can have vacuoles

QUIZ

1. A) mitochondrion, B) nucleus, C) rough endoplasmic reticulum (rough ER), D) Golgi apparatus
2. B
3. A
4. A
5. To contain **DNA/genetic material**; to **control the cell's activities**.
6. To provide **support and structure** for the plant. (It does not have a skeleton.)
7. To **make proteins** for the cell.
   You can also say "synthesize" instead of "make" — this is great scientific vocabulary!

Specialized Cells *pp. 48–53*

ACTIVATE

1. Labels (next page):

© Emma The Teachie Limited 2025-Present. All rights reserved.

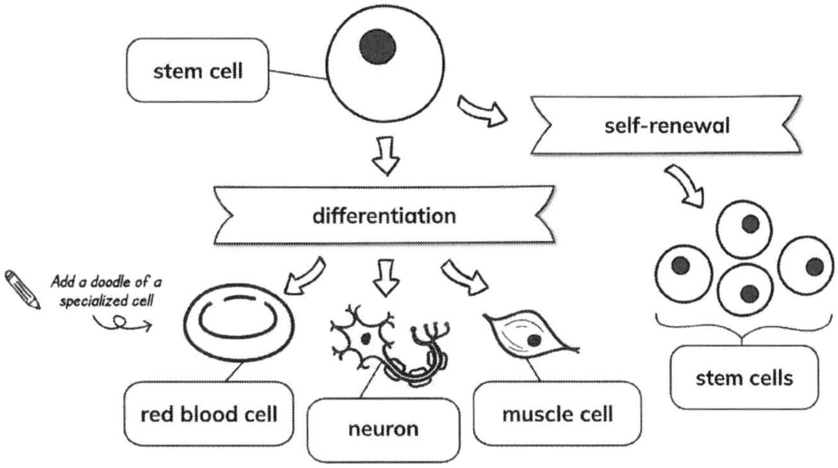

2. Match-up:

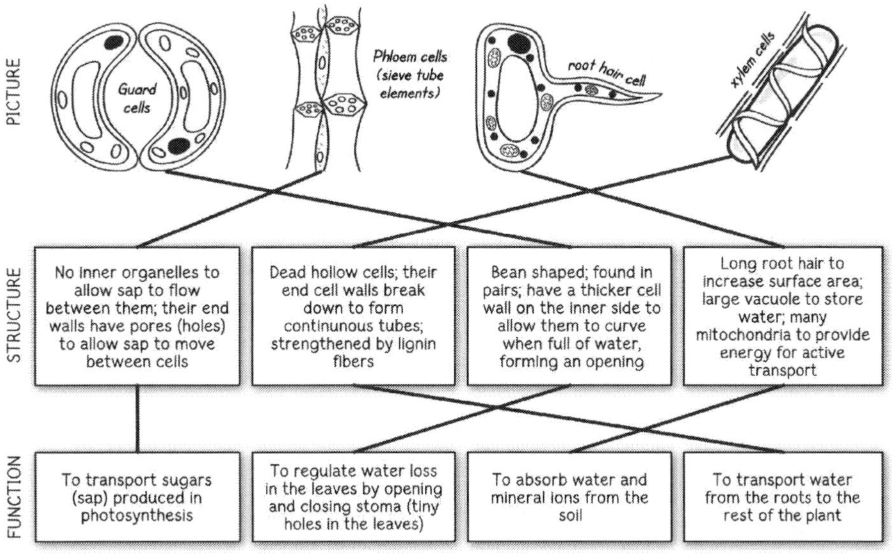

3. For every structure you mention, you must explain how it helps it carry out its function: "The sperm has X to allow it to Y."
   The sperm has a long tail to allow it to swim through the female reproductive system.
   It has many mitochondria in its midpiece to release energy for swimming.
   The head has a pointed shape to make penetration of the egg easier.
   The acrosome in the head contains digestive enzymes to break down the egg's protective outer layers.
   The head contains a nucleus with DNA to allow it to combine with the egg cell's DNA.

4. Axons; hemoglobin; surface area; mitochondria; vacuole; filaments; energy.

EXTEND

1. Blood cells

2. Some of the donor cells will differentiate into new blood cells, and some of them will self-renew to regenerate healthy bone marrow tissue.

REVIEW

1. Neuron
2. Specialized
3. Root hair cell

© Emma The Teachie Limited 2025-Present. All rights reserved.

D. Function
E. Stem cell
F. Red blood cell

QUIZ

1. A) root hair cell, B) red blood cell, C) neuron/nerve cell, D) sperm cell
2. B
3. B
4. D
5. Differentiation
6. To release energy for swimming
7. They contain **protein filaments** that contract and relax to cause **movement.** They have many **mitochondria** to **release energy** for movement.

The Cell Membrane

pp. 56–61

ACTIVATE

1. Four functions — for each one, you should have a description or doodle:
   Gatekeeper / selective barrier - controls what can enter and exit the cell
   Cell signaling - cells communicating with each other
   Cell recognition - cells recognizing other cells as either “self” or “foreign”
   Cell adhesion - cells sticking together to form tissues and organs

2. Labels (extra possible labels shown in italics):

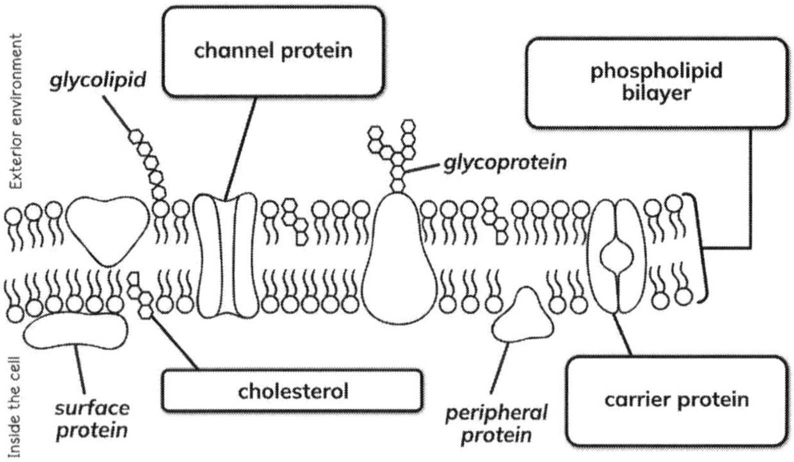

3. Your own storyboard — each image should show what is happening in the description:

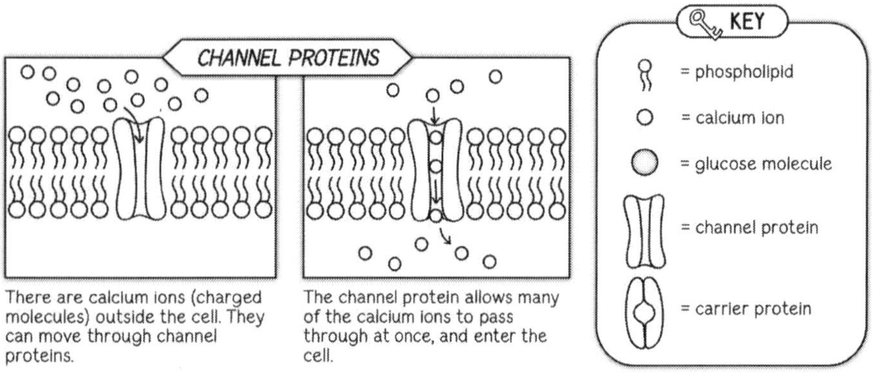

There are calcium ions (charged molecules) outside the cell. They can move through channel proteins.

The channel protein allows many of the calcium ions to pass through at once, and enter the cell.

© Emma The Teachie Limited 2025-Present. All rights reserved.

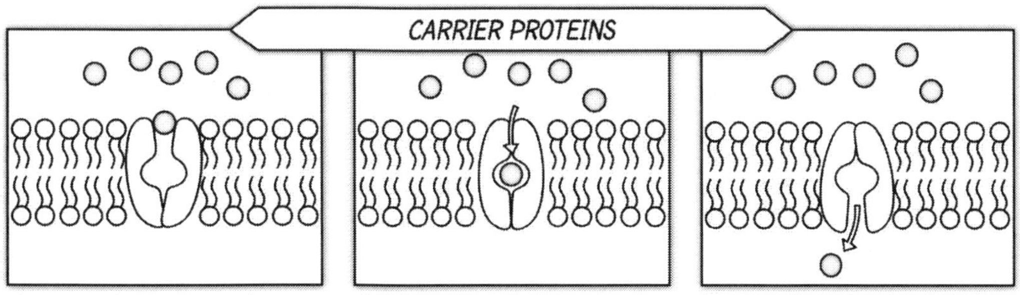

There are glucose molecules outside the cell. One of them binds to the surface of the carrier protein.

This binding causes the carrier protein to change shape, bringing the glucose through the cell membrane.

The glucose molecule is released inside the cell, and the carrier protein returns to its normal shape.

4. Labels: speech bubbles should relate to hydrophilic and hydrophobic properties:

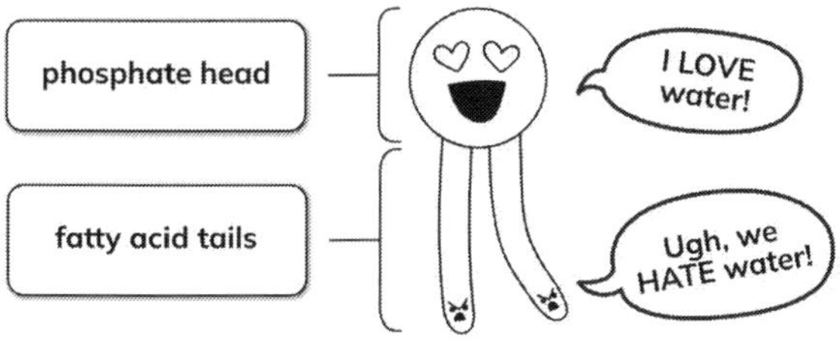

EXTEND

Open ended — your ideas.

Sample answer: They could develop a medicine that detects the "dead cell" molecules and destroy the cells with these. Any real dead cells are already dead, and any viruses displaying these molecules will be destroyed.

REVIEW

A. Bilayer
B. Fluid
C. Nonpolar; proteins
D. Cholesterol
E. Enter; exit (in either order)
F. Recognition

QUIZ

1. A) phospholipid, B) carrier protein, C) cholesterol, D) channel protein
2. A and D
3. D
4. C
5. Cholesterol
6. In a **bilayer**. The hydrophobic **fatty acid tails face inwards**, and the hydrophilic **phosphate heads face outwards**.
7. Channel proteins and carrier proteins

© Emma The Teachie Limited 2025-Present. All rights reserved.

Diffusion and Osmosis

pp. 62–67

ACTIVATE

1. Your own diagrams. They should show the dye initially being concentrated in one area (when first dropped in) and then spreading out until the whole glass of water is the same color.

   Your explanation should state that when the dye is first added, there is a high concentration of dye in one area and a low concentration (no dye) in the rest of the glass of water. The dye moves by diffusion from high to low concentration until equilibrium is achieved and there is an equal concentration of dye everywhere.

2. "Before" diagrams should have an arrow to show net movement of molecules. "After" diagrams must show an equal number of molecules on both sides. All diagrams should have labels:

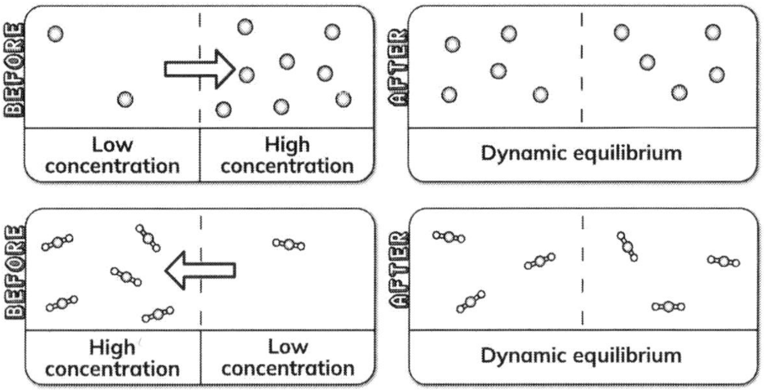

3. Numbers and arrows (and in two cases, you should have written "No net movement."

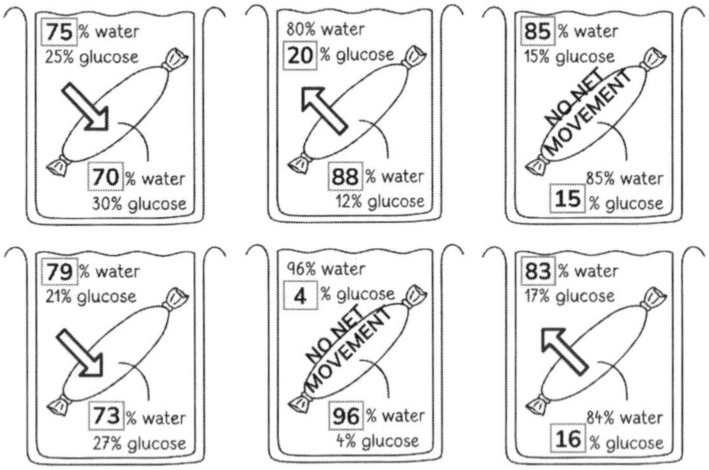

4. Match-up lines in this order:

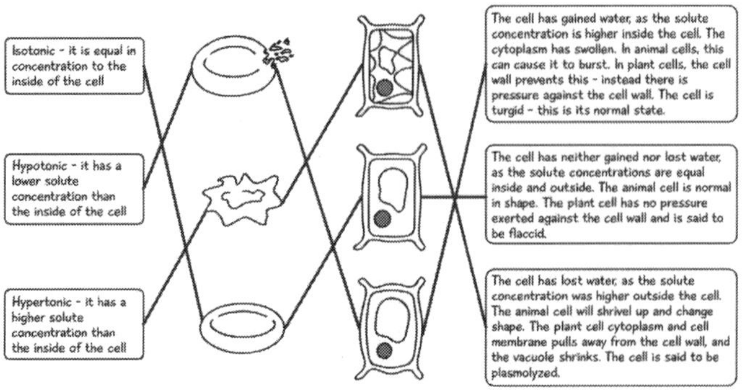

© Emma The Teachie Limited 2025-Present. All rights reserved.

EXTEND

Drinking a lot of water in a short period would lead to a **high concentration of water** in the blood — i.e., a **low solute concentration.** This leads to **water moving by osmosis** out of the blood and into the body's cells, **causing them to swell** and not function properly.

REVIEW

Mistakes you should have underlined:

Diffusion is the process of molecules moving from an area of lower concentration to an area of higher concentration. As it is down a concentration gradient, it is an active process.

When the concentration is equal, the molecules stop moving. This is called dynamic momentum.

Osmosis is a special type of diffusion. It is the movement of water molecules, from a high concentration of water to a low concentration of water, across a permeable membrane.

We can also say that water moves from a low solute concentration to a high solute concentration.

And here are the corrections:

Diffusion is the process of molecules moving from an area of **higher** concentration to an area of **lower** concentration. As it is down a concentration gradient, it is a **passive** process.

When the concentration is equal, the molecules **continue to move**, but there is **no further net movement.** This is called dynamic **equilibrium.**

Osmosis is a special type of diffusion. It is the movement of water molecules from a high concentration of water to a low concentration of water across a **semipermeable** membrane.

We can also say that water moves from a low solute concentration to a high solute concentration.

QUIZ

1. A) →, B) No net movement, C) ←, D) ←
2. A
3. A and D
4. A
5. Equilibrium
6. The cell membrane
7. Water will **move out** of the red blood cell **by osmosis**. The cell will shrink/shrivel up.

Active Transport *pp. 68–73*

ACTIVATE

1. A) passive, B) active, C) passive, D) passive (even though a carrier protein is involved, the molecules are moving from high to low concentration - i.e., down their concentration gradient)

2. Diagram 1: The diagram shows a channel protein, but active transport uses carrier proteins.

   Diagram 2: The diagram shows molecules moving directly through the cell membrane without the use of carrier proteins.

© Emma The Teachie Limited 2025-Present. All rights reserved.

3. Your own diagrams showing the process of endocytosis. They should be labeled and show molecules entering the cell by forming a vesicle made out of the cell membrane:

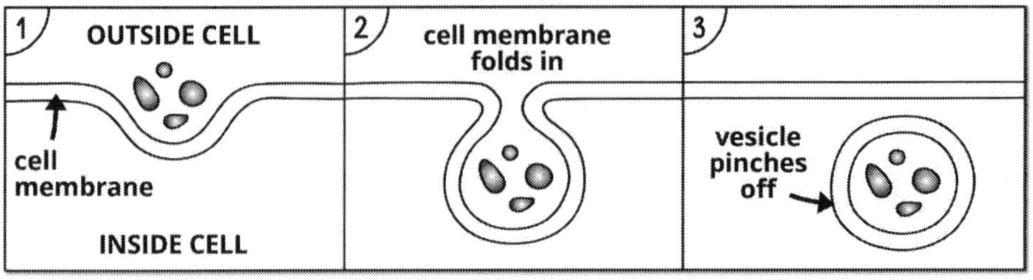

4. Venn diagram:

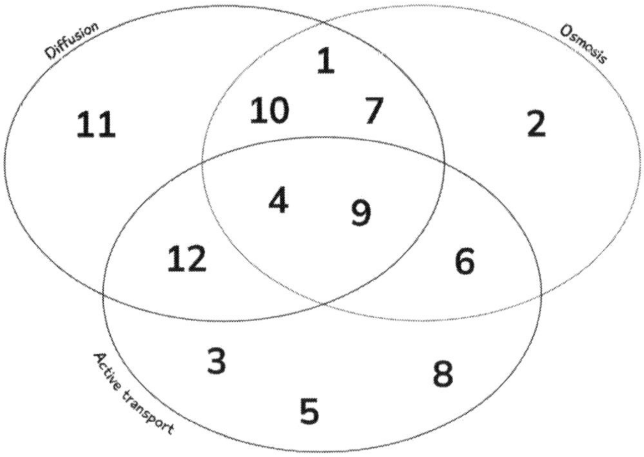

EXTEND

Open response — your own analogy and illustration. It should show that energy is required; it is an active process. It may also illustrate moving up a gradient.

REVIEW

1. True
2. False - it requires extra energy
3. False - it is against a concentration gradient
4. True
5. False - it uses carrier proteins
6. True

QUIZ

1. A) osmosis, B) active transport, C) diffusion, D) diffusion (also known as facilitated diffusion, when carrier proteins are used)
2. A
3. B
4. Osmosis
5. Endocytosis
6. Carrier proteins
7. It moves molecules **against their concentration gradient** from low to high concentration.

© Emma The Teachie Limited 2025-Present. All rights reserved.

ACTIVATE

1. Labels (extra possible labels from "Take it further" task shown in italics):

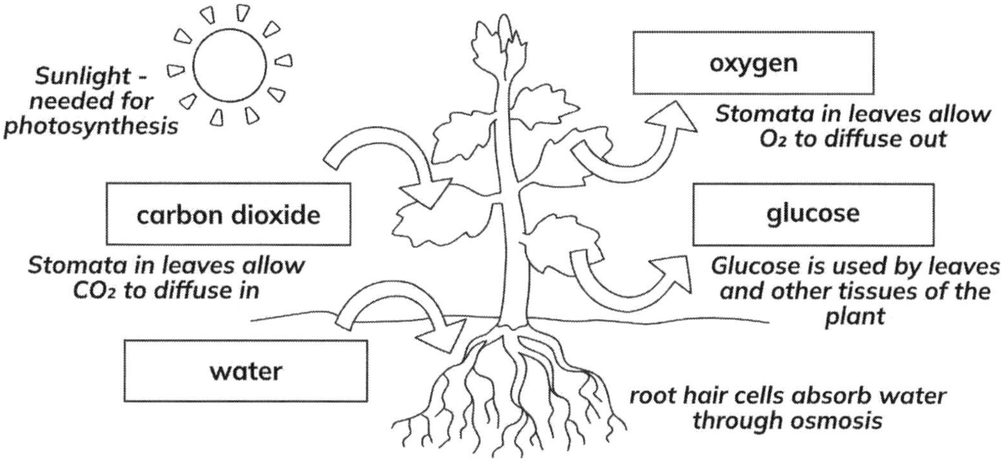

2. Word equation: Carbon dioxide + water → glucose + oxygen (you may have written light along the arrow)
   Symbol equation: $6CO_2 + 6H_2O \rightarrow C_6H_{12}O_6 + 6O_2$

3. What does it transport? Xylem - water; Phloem - glucose (sugar)
   In which directions can it transport molecules? Xylem - upwards only; Phloem - both directions
   Does it require energy to transport molecules? Xylem - no; Phloem - yes

4. A) osmosis; roots; area
   B) transported; xylem
   C) dioxide; stomata
   D) chloroplasts; chlorophyll; glucose; oxygen (glucose and oxygen can be in either order)
   E) phloem; energy
   F) diffuse

5. Check marks in this order:

☑	☐
☐	☑
☐	☑
☑	☐
☑	☐

EXTEND

1. The purpose of stomata in photosynthesis is to allow gas exchange to take place and to regulate water loss from the leaves (also known as transpiration).

2. Stomata are able to close to prevent too much water being lost from the plant — this would lead to it wilting and potentially dying.

3. Stomata are open during the day to allow gas exchange needed for photosynthesis. They are closed at night, as there is no sunlight for photosynthesis. Being closed prevents water from being lost.

© Emma The Teachie Limited 2025-Present. All rights reserved.

REVIEW

Verbal review. All keywords should be checked.

QUIZ

1. A) stoma, B) chloroplast, C) xylem, D) glucose
2. A
3. A
4. B
5. Carbon dioxide + oxygen → glucose + oxygen (you may have written light along the arrow)
6. Chlorophyll
7. **Stomata** in the leaves allow **carbon dioxide** to **diffuse** into the plant.

Cellular Respiration *pp. 80–85*

ACTIVATE

1. Aerobic respiration: glucose + oxygen → carbon dioxide + water
   Anaerobic respiration in animals: glucose → lactic acid
   Anaerobic respiration in yeast: glucose → ethanol + carbon dioxide

2. Answers:

	AEROBIC RESPIRATION IN ANIMAL CELLS	*AEROBIC RESPIRATION IN PLANT CELLS*
Where it happens	Mitochondria	Mitochondria
How the glucose is obtained	Eating / feeding	Photosynthesis
How oxygen is obtained	Breathing in	From photosynthesis or gas exchange in leaves
Products	Carbon dixoide + water	Carbon dixoide + water
What happens to the water produced	Used in cells or released through sweat, urine, breath	Used for photosynthesis or lost through leaves
What happens to the carbon dioxide produced	Transported to lungs and breathed out	Used for photosynthesis or else released through leaves

3. Venn diagram:

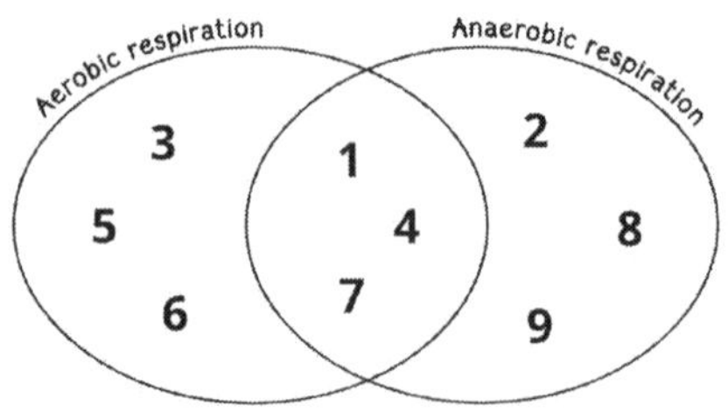

4. Check marks:

STATEMENT	PHOTOSYNTHESIS	AEROBIC RESPIRATION	NEITHER!
Occurs in chloroplasts	✓	○	○
Occurs in nucleus	○	○	✓
Requires oxygen	○	✓	○

© Emma The Teachie Limited 2025-Present. All rights reserved.

Generates ATP/energy	○	✓	○
Happens all the time	○	✓	○
Occurs in mitochondria	○	✓	○
Happens during daylight	✓	✓	○
Requires light energy	✓	○	○
Produces lactic acid	○	○	✓
Produces glucose and oxygen	✓	○	○
Produces water and carbon dioxide	○	✓	○
Done by almost all cells	○	✓	○
Only done by autotrophs	✓	○	○

EXTEND

1. Anaerobic respiration - because there is not enough oxygen.
2. Lactic acid builds up. This causes fatigue and can lead to muscle cramps.
3. The lack of oxygen means glucose is not fully broken down, so there isn't enough energy to sustain sprinting for long periods. Additionally, the lack of oxygen and build up of lactic acid would be harmful to cells.

REVIEW

Open ended — here are some sample questions:

Mitochondria: Where does aerobic cellular respiration take place in eukaryotic cells?
Ethanol: What alcohol is produced during anaerobic respiration in yeast?
ATP: In what form is energy released during cellular respiration?
Glucose: Which sugar is broken down during cellular respiration?
Red blood cells: Which cells transport gases around the human body?
Carbon dioxide: Which gas is produced during aerobic respiration?
Aerobic: Which type of cellular respiration uses oxygen?

QUIZ

1. A) product, B) reactant, C) product, D) reactant
2. B
3. D
4. C
5. Stoma (singular) / Stomata (plural)
6. Any one from:
   - Sweating
   - Urinating
   - Breathing it out
7. Any two from:
   - Photosynthesis requires light; respiration doesn't
   - Photosynthesis occurs in the chloroplasts; respiration occurs in the mitochondria
   - Photosynthesis produces glucose and oxygen; respiration produces CO_2 and water
   - Photosynthesis uses CO_2 and water; respiration uses glucose and oxygen

© Emma The Teachie Limited 2025-Present. All rights reserved.

The Cell Cycle *pp. 88–93*

ACTIVATE

1. Each stage should have doodles and descriptions as follows:
   G0: resting phase, unfavorable conditions for growth (doodles are already done for G0)
   G1: new proteins and organelles synthesized (doodles should show new organelles)
   S: DNA is replicated (doodles should show this in the nucleus)
   G2: proteins for cell division made - centrosomes and spindle fibers (doodles to show two centrosomes)
   M / Mitosis: the two copies of the DNA are separated and the cell divides (doodles are already done for mitosis)

2. Order of check marks:

☑	☐
☑	☐
☐	☑
☑	☐
☐	☑

3. Order: 3; 2; 4; 1. Phase: G2; S; Mitosis; G1

4. These are your own definitions, so they will not exactly match. Add any extra bits you didn't have in your definitions:

   Interphase: A phase of the cell cycle when the cell grows and prepares for cell division
   Mitosis: The process of cell division in eukaryotic cells
   Replicate: To make another copy of; in the cell cycle, the DNA replicates during "S" phase
   Centrosome: Made from two centrioles, the centrosome is important for mitosis

EXTEND

1. The cell will not continue through the cell cycle, and it may be destroyed (apoptosis).

2. Cancer cells have mutations that cause them to divide uncontrollably. The cell cycle is not regulated.

3. Chemotherapy targets rapidly dividing cells. This will attack and kill the cancer cells. However, hair follicles are also rapidly dividing, so these may be damaged as a side effect, causing hair loss.

REVIEW

Open ended — you should have some notes added here.

QUIZ

1. A) S (two copies of DNA visible), B) G2 (two centrosomes visible), C) G1 (new organelles visible), D) G0 (no preparation for cell division happening)
2. C
3. A
4. B
5. Two genetically identical cells
6. New organelles and proteins are made
7. During cell division, the **DNA is separated into two cells**, so it must be copied before this.

© Emma The Teachie Limited 2025-Present. All rights reserved.

Mitosis pp. 94–99

ACTIVATE

1. Repair, growth, and asexual reproduction
2. Open ended — your own mnemonic
3. A) Chromosome, B) Chromosome, C) Chromatid, D) Chromosome
4. Your own diagrams — these should be similar to the ones shown in the text passage. Labels should describe what is happening.

EXTEND

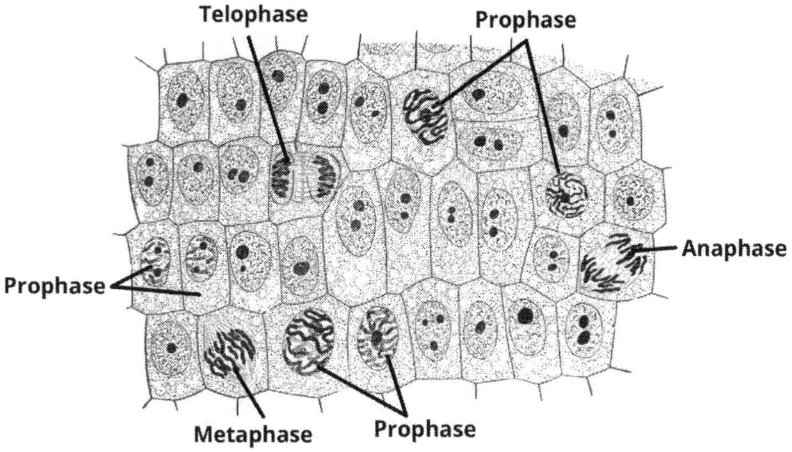

Mitotic index = cells with visible chromosomes / total number of cells in the field of view
Mitotic index = 9 / 34
Mitotic index = 0.26 (to 2 decimal points)

REVIEW

1. Metaphase
2. Telophase
3. Prophase
4. Anaphase
5. Metaphase
6. Anaphase
7. Prophase

QUIZ

1. A) metaphase, B) telophase, C) prophase, D) anaphase
2. C
3. D
4. A
5. Prophase → Metaphase → Anaphase → Telophase
6. Cytokinesis is when the cell membrane pinches in and divides the cell in two. It happens after telophase.
7. Growth, repair, and asexual reproduction

© Emma The Teachie Limited 2025-Present. All rights reserved.

Meiosis

pp. 100–105

ACTIVATE

1. Haploid - 3; Diploid - 5; Crossing over - 6; Cytokinesis - 1; Gametes - 4; Homologous chromosomes - 2.

2. A) Two
   B) Crossing over
   C) Anaphase I
   D) It is not preparing to divide, so there is just one of each chromosome.

3. Cell division; Gametes; Replicated; Break down; Haploid; Unique; Fertilization; Chromosomes; Diploid

4. Venn diagram:

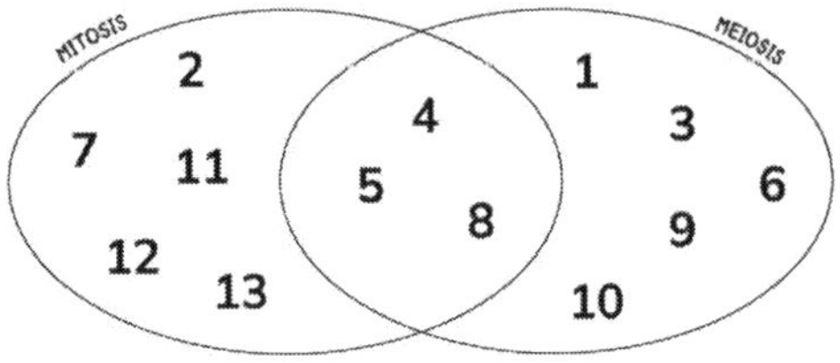

EXTEND

8
32
44
19
64
229
78

Your own answer — you may have been surprised that the Atlas blue butterfly has such a high number of chromosomes, or that the horse and guinea pig have the same number of chromosomes!

Chromosome number does not indicate complexity of an organism; there is no correlation.

REVIEW

False - Meiosis results in haploid cells.
True
False - Human gametes contain 23 chromosomes.
True
False - Meiosis makes gametes.

QUIZ

1. A) telophase, I B) telophase II, C) metaphase I, D) anaphase II
2. D
3. D
4. A
5. Four
6. Haploid = one set of chromosomes; Diploid = two sets of chromosomes.
7. The homologous chromosomes are **separated** to **opposite poles** of the cell by the **contraction of spindle fibers**.

© Emma The Teachie Limited 2025-Present. All rights reserved.

Sexual and Asexual Reproduction

pp. 106–111

ACTIVATE

1. Your own definitions.
   Sexual reproduction: Two parents combine DNA to produce genetically unique offspring.
   Asexual reproduction: One parent produces offspring that are genetically identical to the parent.

2. Advantages of sexual reproduction:
   - Offspring more likely to survive new diseases and in new environments
   - Harmful mutations can be removed from population

   Advantages of asexual reproduction:
   - Quicker
   - Can produce many offspring

   Disadvantages of sexual reproduction:
   - Produces fewer offspring
   - Less energy efficient/slower (have to find a mate)

   Disadvantages of asexual reproduction:
   - Any mutations always passed on
   - Offspring less likely to be adapted to new environments/new diseases

3. Match up:

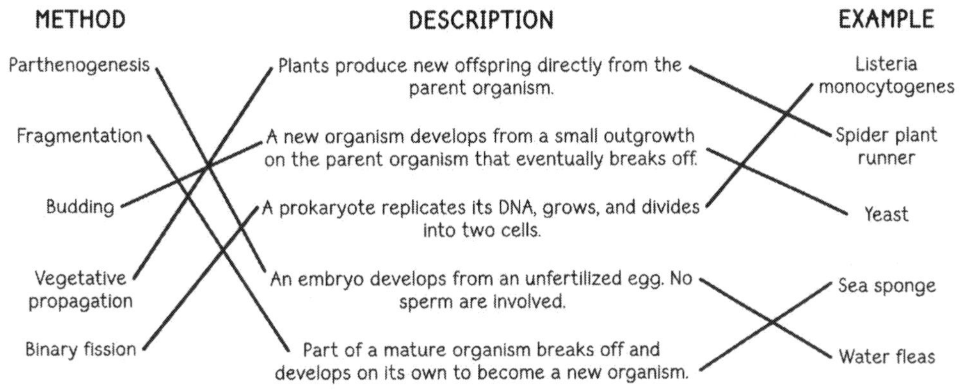

4. A) The "X"s should be drawn over any beetles that are a different color than the leaf. And your answer is: These beetles wouldn't survive because they are not camouflaged, so predators would eat them.

   B) All of the beetles would be green. All beetles on leaf 1 would survive. No beetles on leaf 2 would survive.

   C) Sexual reproduction produces variation in offspring (asexual reproduction produces clones), which increases their chances of survival in a changing environment, as they may have adaptations.

EXTEND

Open ended — you should have a drawing of your organism and answered each question:
What environment it lives in (your organism should show some adaptations to suit this environment);
When it uses asexual reproduction (when conditions are favorable); and
When it switches to sexual reproduction (when conditions are unfavorable/changing rapidly).

© Emma The Teachie Limited 2025-Present. All rights reserved.

REVIEW

Mistakes you should have underlined:

Sexual reproduction involves two parents. It is when two body cells fuse together, combining the RNA. This creates genetic speciation in the offspring.

Asexual reproduction involves one or two parents. The offspring are genetic twins of the parent.

Sexual reproduction is faster than asexual reproduction, and the variation in offspring gives them a greater chance of surviving in a new environment. Asexual reproduction is slower but allows more offspring to be produced if the conditions are favorable.

And here are the corrections:

Sexual reproduction involves two parents. It is when two **gametes** fuse together, combining their **DNA**. This creates genetic variation in the offspring.

Asexual reproduction involves **one parent only**. The offspring are genetic **clones** of the parent.

Sexual reproduction is **slower** than asexual reproduction, and the variation in offspring gives them a greater chance of surviving in a new environment. Asexual reproduction is **faster** but allows more offspring to be produced if the conditions are favorable.

QUIZ

1. A) vegetative propagation, B) fragmentation, C) parthenogenesis, D) binary fission
2. A
3. B
4. B
5. Any one:
   It is quicker;
   It is more energy efficient;
   It is possible when an individual is isolated; or
   It can produce many offspring.
6. When one parent produces offspring that are **genetically identical to the parent**.
7. Sexual reproduction produces **genetic variation/diversity** in offspring, so they are more likely to have **adaptations** that allow them to survive a **changing environment/new disease**.

YOU'VE FINISHED THE LESSONS!!!

That's amazing! Think of everything you have learned — it is a lot!

There is a whole book assessment included in this workbook.

BEFORE you complete it, you need to study for it.

Choose some ways to study: reread text passages • create flashcards • explain a lesson to someone • do the lesson quizzes again (use paper/card to cover up your original answers) • make summary posters • do practice questions • mind dump (write down everything you know about a lesson, then add any missed parts from your notes/text passages)

© Emma The Teachie Limited 2025-Present. All rights reserved.

Whole Book Assessment

INSTRUCTIONS: It is recommended to remove these pages from the book so you do not have access to the rest of the workbook while completing the test.

Complete in a quiet space with no disturbances. Use a pencil and eraser.

There are 55 multiple choice questions. Choose one answer for each question. Fill in the circle beside your chosen answer. If you change your mind, fully erase your pencil mark.

TIME: You have 40 minutes to complete the test.

1. In bacteria, DNA is located in the:

   - O Ribosome
   - O Nucleus
   - O Cytoplasm
   - O Mitochondria

2. A class collects plant growth data for two weeks. What should they do next?

   - O Form a hypothesis
   - O Record conclusions
   - O Analyze the data for patterns
   - O Change their variables

3. What is formed when two gametes combine?

   - O A clone
   - O A diploid cell
   - O A haploid cell
   - O A bud

4. Microscopes helped scientists develop cell theory because they:

   - O Proved spontaneous generation
   - O Created cell structures
   - O Replaced experiments
   - O Made cells visible for study

5. An organism eats, moves, and reproduces. Which other trait shows that it is alive?

   - O It grows
   - O It's colorful
   - O It stays in water
   - O It's visible

6. Which type of molecules can pass directly through the cell membrane?

   - O Small polar molecules
   - O Small nonpolar molecules
   - O Large polar molecules
   - O Large nonpolar molecules

7. What happens before a cell divides?

   - O DNA is copied
   - O Organelles vanish
   - O Membrane thickens
   - O ATP breaks down

8. Red blood cells lack nuclei to:

   - O Fight infections
   - O Divide more quickly
   - O Absorb nutrients
   - O Carry more oxygen efficiently

9. Which list shows levels from smallest to largest?

   - O Cell → Tissue → Organ → Organism → Organ system
   - O Cell → Tissue → Organ → Organ system → Organism
   - O Organ → Cell → Tissue → Organ system → Organism
   - O Organism → Organ → Cell → Tissue → Organ system

© Emma The Teachie Limited 2025-Present. All rights reserved.

10. What term means the membrane allows only some materials through?

- Permeable
- Transparent
- Semipermeable
- Flexible

11. What is made during photosynthesis?

- Oxygen and glucose
- Carbon dioxide and water
- Water and glucose
- Oxygen and water

12. Cellular respiration mainly takes place in the:

- Mitochondria
- Chloroplasts
- Nucleus
- Ribosome

13. Which sequence is correct?

- Anaphase → Telophase → Prophase → Metaphase
- Metaphase → Prophase → Telophase → Anaphase
- Prophase → Metaphase → Anaphase → Telophase
- Telophase → Prophase → Metaphase → Anaphase

14. Which example shows exocytosis?

- Protein entering cell
- Cell releasing waste via vesicle
- Water moving into cell
- Oxygen diffusing

15. Which process describes a stem cell becoming specialized?

- Respiration
- Diffusion
- Homogenization
- Differentiation

16. What is the purpose of a control in an experiment?

- To test more variables
- To compare results
- To make data random
- To increase bias

17. Which pair is correctly matched?

- Ribosome – energy storage
- Mitochondria – protein synthesis
- Chloroplast – photosynthesis
- Vacuole – genetic control

18. Phospholipids are part of the cell membrane. Which part of a phospholipid is hydrophobic?

- Nucleus
- Fatty acid tails
- The glycoprotein
- Phosphate head

19. During which phase do chromosomes line up in the middle of the cell?

- Prophase
- Anaphase
- Telophase
- Metaphase

20. Which of these would not be found in a prokaryotic cell?

- Nucleus
- Ribosome
- Cytoplasm
- Cell wall

21. In cell theory, the botanist Matthias Schleiden concluded that:

- Cells arise from nonliving matter
- All animals are made of cells
- All plants are made of cells
- Cells can live outside organisms

© Emma The Teachie Limited 2025-Present. All rights reserved.

22. Which example shows the correct order of ecological organization?

- O Ecosystem → Biosphere → Population → Community
- O Population → Community → Ecosystem → Biosphere
- O Community → Ecosystem → Biosphere → Population
- O Organism → Biosphere → Ecosystem → Community

23. What is the function of photosynthesis?

- O To transport molecules around the plant
- O To produce food for the plant
- O To absorb water for the plant
- O To absorb light for the plant

24. Meiosis occurs in:

- O Skin for healing
- O Brain tissue for growth
- O Muscles for repair
- O Sex organs for gamete production

25. Which process is the opposite of diffusion?

- O Passive flow
- O Osmosis
- O Homeostasis
- O Active transport

26. Which cell structure is found in both animal and plant cells?

- O Chloroplast
- O Lysosome
- O Cell wall
- O Rough endoplasmic reticulum

27. Homeostasis helps organisms to:

- O Maintain stable internal conditions
- O Absorb energy from the sun
- O Grow faster in heat
- O Change their DNA

28. Which part of a microscope is the slide placed on?

- O The base
- O The stage
- O The eye piece
- O The objective lens

29. How does water move in osmosis?

- O From inside the cell to outside the cell
- O From high solute to low solute concentration
- O From high water to low water concentration
- O From low water to high water concentration

30. Which setup shows diffusion?

- O Boiling water
- O Food coloring spreading in water
- O Ice melting in sunlight
- O Mixing sugar with a spoon

31. Which structure is found in all cells?

- O Cell membrane
- O Chloroplast
- O Mitochondria
- O Nucleus

32. In cell division, which structure is responsible for separating chromosomes?

- O Homologous chromosomes
- O Centrioles
- O Nuclear envelope
- O Spindle fiber

33. How can we describe the cell membrane based on its properties?

- O Rigid wall
- O Permeable bilayer
- O Impenetrable barrier
- O Fluid mosaic

© Emma The Teachie Limited 2025-Present. All rights reserved.

34. The waste products of cellular respiration are:

- O ATP and light
- O Oxygen and glucose
- O Carbon dioxide and water
- O Sugar and starch

35. Crossing over during meiosis increases:

- O Genetic variation
- O Mutation rate
- O Chromosome number
- O Cell energy

36. Fragmentation is one form of asexual reproduction. Which is another?

- O Fertilization
- O Budding in yeast
- O Meiosis in animals
- O Pollination

37. Which variable is changed in experiments to see the effect of it?

- O Independent
- O Dependent
- O Control
- O Fixed

38. The lungs and heart working together are part of which level?

- O Tissue
- O Organism
- O Organ
- O Organ system

39. Muscle cells have many mitochondria because they:

- O Need energy for contraction
- O Make DNA
- O Store waste
- O Transport oxygen

40. A red blood cell is placed into a beaker of pure water. What happens to it over time?

- O It becomes wrinkled and nonfunctional
- O It grows in size and may burst
- O It is unaffected and functions normally
- O It loses water and shrinks in size

41. In the cell cycle, why must DNA replicate before mitosis?

- O So each new cell has a full set of chromosomes
- O To form gametes
- O To make energy
- O To repair enzymes

42. What happens in cells when oxygen is low?

- O Photosynthesis begins
- O Anaerobic respiration occurs
- O ATP production stops fully
- O The cell dies instantly

43. Mitosis produces:

- O Four haploid cells
- O A single large cell
- O Two identical diploid cells
- O Many different cells

44. Binary fission occurs in:

- O Humans
- O Insects
- O Bacteria
- O Plants

45. Why is extra energy needed in active transport?

- O To cool the cell down while it is working
- O To move the molecules quickly enough
- O To move molecules against their concentration gradient
- O To maintain equilibrium and stop the molecules moving

© Emma The Teachie Limited 2025-Present. All rights reserved.

46. Which process happens only in meiosis I?

- O Nucleus reforms
- O Cytoplasm divides
- O Homologous chromosomes pair up
- O Chromatids separate

47. Which of these is not a component of cell theory?

- O All living things are made of cells
- O Cells can form from nonliving matter
- O The cell is the basic unit of life
- O All cells come from preexisting cells

48. A lizard moves into the shade to cool down. What characteristic is shown by this?

- O Growth
- O Response to stimuli
- O Energy use
- O Adaptation

49. Which statement best describes eukaryotic cells?

- O They only exist in bacteria
- O They lack DNA
- O They are smaller than prokaryotes
- O They have a nucleus and organelles

50. What is the function of lysosomes?

- O Store energy
- O Control cell division
- O Contain digestive enzymes to break down waste
- O Make lipids

51. Which pigment absorbs light for photosynthesis?

- O Chloroplast
- O Chlorine
- O Chlorophyll
- O Hemoglobin

52. Order the stages of the cell cycle correctly:

- O $G_2 \rightarrow G_1 \rightarrow$ S phase
- O Cytokinesis → Mitosis → Interphase
- O Interphase → Mitosis → Cytokinesis
- O Mitosis → Interphase → Cytokinesis

53. What happens to DNA during prophase?

- O It decondenses into chromatin
- O It replicates to make two copies
- O It gets broken down with the nuclear envelope
- O It condenses into chromosomes

54. Sexual reproduction increases survival because:

- O Offspring are genetically diverse
- O It happens faster
- O It avoids meiosis
- O All offspring are identical

55. When perfume spreads across a room, it's an example of:

- O Diffusion
- O Osmosis
- O Active transport
- O Evaporation

FINISHED?

If you have spare time, check your answers.

If you have left any blank, make an educated guess.

© Emma The Teachie Limited 2025-Present. All rights reserved.

Whole Book Assessment - Reflection

Well done on completing the assessment!

After you have marked it, it is beneficial to reflect on your learning. Science has shown that repeated self-reflection after testing can improve future performance! Let's do that by answering these questions:

My score: _____/______. Before seeing results, I predicted: _____/______.

How close were you?

Pick three questions you got incorrect. For each, circle the main cause and explain.

Question 1: misread/skimmed it • gap in knowledge • unknown vocabulary • didn't review this material • ran out of time • careless/other: ______________________

Explanation: ____

Question 2: misread/skimmed it • gap in knowledge • unknown vocabulary • didn't review this material • ran out of time • careless/other: ______________________

Explanation: ____

Question 3: misread/skimmed it • gap in knowledge • unknown vocabulary • didn't review this material • ran out of time • careless/other: ______________________

Explanation: ____

Which study techniques did you use before the test? Circle the main ones

Practice questions • self-testing/flashcards • rereading notes • explaining to someone • spacing practice over days • cramming night before • other:

Which of these helped the most/least? Why?______________________________

____

Two specific actions I will take before future assessments:

Action 1 (what, when, where, how long): ______________________________

____

Action 2 (what, when, where, how long): ______________________________

____

Accountability: share these actions with someone else and ask them to check in that you have done them by when you said you would!

© Emma The Teachie Limited 2025-Present. All rights reserved.

Whole Book Assessment - ANSWER KEY

1. ○○●○
2. ○○●○
3. ○●○○
4. ○○○●
5. ●○○○
6. ○●○○
7. ●○○○
8. ○○○●
9. ○●○○
10. ○○●○
11. ●○○○
12. ●○○○
13. ○○●○
14. ○●○○
15. ○○○●
16. ○●○○
17. ○○●○
18. ○●○○
19. ○○○●
20. ●○○○
21. ○○●○
22. ○●○○
23. ○●○○
24. ○○○●
25. ○○○●
26. ○○○●
27. ●○○○
28. ○●○○
29. ○○●○
30. ○●○○
31. ●○○○
32. ○○○●
33. ○○○●
34. ○○●○
35. ●○○○
36. ○●○○
37. ●○○○
38. ○○○●
39. ●○○○
40. ○●○○
41. ●○○○
42. ○●○○
43. ○○●○
44. ○○●○
45. ○○●○
46. ○○●○
47. ○●○○
48. ○●○○
49. ○○○●
50. ○○●○
51. ○○●○
52. ○○●○
53. ○○○●
54. ●○○○
55. ●○○○

Once you've marked your assessment, complete the self-reflection questions.

© Emma The Teachie Limited 2025-Present. All rights reserved.

© Emma The Teachie Limited 2025-Present. All rights reserved.

Want a free workbook bonus?

Scan the QR code with your device to find out what it is!

Share your experience

Did your child enjoy this workbook? Did they have a favorite lesson/activity? If so, I'd love to hear all about it!

Your review helps other families to find workbooks that meet their needs.

I read and appreciate every review!

Thank you,

Emma

View more of my workbooks!

Copyrighted Materials: All rights reserved.
© Emma The Teachie Limited 2025

Contact the author:
Emmatheteachie.com

Made in United States
Orlando, FL
10 November 2026

90510176R00081